Distributed CMOS Bidirectional Amplifiers

ANALOG CIRCUITS AND SIGNAL PROCESSING

Series Editors:
Mohammed Ismail, *The Ohio State University*
Mohamad Sawan, *École Polytechnique de Montréal*

For further volumes:
http://www.springer.com/series/7381

Ziad El-Khatib • Leonard MacEachern
Samy A. Mahmoud

Distributed CMOS Bidirectional Amplifiers

Broadbanding and Linearization Techniques

 Springer

Ziad El-Khatib
Department of Electronics
Carleton University
1125 Colonel By Drive
Ottawa, ON
Canada

Leonard MacEachern
Department of Electronics
Carleton University
1125 Colonel By Drive
Ottawa, ON
Canada

Samy A. Mahmoud
Department of Systems
and Computer Engineering
Carleton University
1125 Colonel By Drive
Ottawa, ON
Canada

ISBN 978-1-4899-8656-6   ISBN 978-1-4614-0272-5 (eBook)
DOI 10.1007/978-1-4614-0272-5
Springer New York Heidelberg Dordrecht London

*To my dear mother and family and friends
for their love and kindness*

Preface

Active MOSFET devices can be modeled by nonlinear current and charge sources that depend on the device voltages. These nonlinear sources give rise to distortion when driven with a modulated signal. When the input signal driven into the amplifier semiconductor is increased, the output is also increased until a point where distortion products can no longer be ignored. The harmonics and higher order distortion of the output signal are generated by nonlinearities of MOSFET devices.

A highly-linear optical transmitter with fully-integrated broadband design linearization capability is required to address linearity improvements. In response to the need to correct the broadband distributed amplifier (DA)'s nonlinear distortion, a number of DA linearization techniques have been developed. However, most of the published DA linearization methods reported do not provide fully-integrated distortion cancellation techniques with large third-order intermodulation (IM3) distortion reduction.

In this book, we demonstrate a fully-integrated fully-differential linearized CMOS distributed bidirectional amplifier that achieves large IMD3 distortion reduction over broadband frequency range for both RF paths. The proposed linearized bidirectional DA has the drain and gate transmission-lines stagger-compensated. Reducing the DA IM3 distortion by mismatching the gate and drain LC delay-line ladders. The proposed fully-differential linearized DA employs a cross-coupled compensator transconductor to enhance the linearity of the DA gain cell with a nonlinear drain capacitance compensator for wider linearization bandwidth. The proposed linearized CMOS bidirectional DA achieves a measured IM3 distortion reduction of 20 dB with frequency of operation from 0.1 to 9.5 GHz and a two-way amplification of 5 dB in both RF directions. The proposed linearized DA is implemented in 0.13 μm RF CMOS process for use in highly-linear UWB communications.

Preface

Acknowledgements

I would like to express my sincere gratitude to my dear co-authors and advisors Dr. Samy A. Mahmoud, Professor at the Faculty of Systems Engineering and Design at Carleton University and Dr. Leonard MacEachern Professor at Electronics and Electrical Engineering at Carleton University for their continuous support, guidance, encouragement and involvement during the course of this book. Their help and support is greatly appreciated.

I am in debt to all who provided support. Financial support was provided by Carleton University in the form of a research assistantship and in the form of a graduate scholarship, and by the National Capital Institute of Telecommunications (NCIT) and Centre for Photonics Fabrication Research (CPFR) in the form of a research grant is gratefully acknowledged. I would like to thank the Canadian Microelectronics Corporation (CMC) staff for both Cadence technical support and fabrication silicon space. I would like to thank IBM Corporation MOSIS for chip fabrication and Agilent staff for their ADS design environment tools and test equipment support to measure the chip performance.

I wish to thank my dear mother, my sisters and brothers for their continuous prayers, encouragement and support during the course of this book, they all have stood by me.

Contents

List of Figures

List of Tables

List of Tables

List of Symbols

ACPR	Adjacent Channel Power Ratio
A_d	Drain line attenuation per section
A_g	Gate line attenuation per section
A_v	Voltage gain
ADS	Advanced Design System
ASITIC	Analysis and Simulation of Spiral Inductors and Transformers for IC
BER	Bit Error Rate
CDMA	Code Division Multiple Access
C_{gs}	Gate-source capacitance
C_{gd}	Gate-drain capacitance
C_{ox}	Capacitance of oxide
C_{si}	Capacitance of silicon substrate
CMOS	Complementary Metal Oxide Semiconductor
dB_c	decibel relative to a carrier level
DA	Distributed Amplifier
dB	Decibel
dBm	Decibel referenced to 1 mW
DQPSK	Differential Quadrature Phase Shift Modulation
EM	Electromagnetic
EVM	Error Vector Magnitude
ESD	Electrostatic Discharge
ε_r	Relative dielectric constant of substrate
f_{max}	Maximum frequency of oscillation
f	Frequency in Hertz
f_{res}	Resonance Frequency
f_T	Unity current-gain frequency
GPIB	General Purpose Interface Bus
g_m	Transconductance
GBW	Gain-bandwidth product
HFSS	High Frequency Structure Simulator

IM3	Third-order (in-band) intermodulation distortion
IIP3	Input Third-order intercept point
IBO	Input Back-off
IM	Intermodulation
IMD	intermodulation distortion
I_d	Drain current
IM/DD	Intensity Modulation Direct Detection
IF	Intermediate Frequency
L_d	Output drain line inductance
L_g	Input gate line inductance
Mbps	Mega Bits Per Second
MGTR	Multi-Gate Transistor
MIM	Metal-Insulator-Metal
NADC	North American Digital Cellular
N	Number of DA Gain Stages
N_{opt}	Optimum number of gain stages
NF	Noise Figure
OFDM	Orthogonal Frequency Division Multiplexing
OIP3	Output Third-order intercept point
PAR	Peak to Average Power Ratio
PSD	Power Spectral Density
P_{1dB}	1dB Compression Point
P_{out}	Output power
QAM	Quadrature Amplitude Modulation
Q	Quality factor
RFPA	RF Power Amplifier
RF	Radio Frequency
R_{si}	Resistance of silicon substrate
RF	Radio Frequency
RoF	Radio-over-Fiber
SRF	Self-resonant frequency
THD	Total harmonic distortion
UWB	Ultra wideband
ω_c	Cut-off frequency
Z_I	Image impedance
Z_o	Transmission line characteristic impedance

Chapter 1
Introduction

1.1 Motivation

RF photonics technology such as Radio-over-fiber (RoF) is a way to distribute RF signals to base stations and Cable TV networks [3–5]. In RoF direct modulation transmission systems, the input current of a semiconductor laser is directly modulated by the information-bearing RF signal. Due to the nonlinear relationship between the input current and the output optical power of the semiconductor laser, distortion is introduced to the RoF transmission system.

Transmitter nonlinearity produces harmonics and intermodulation distortion (IMD) products. Some of these products fall within the transmission band and can degrade the RoF system performance. For good service quality, it is required to keep the distortion below a certain level [6–8]. Cable TV networks requires a carrier to distortion ratio better than 50 dB [6–8]. Other wireless services such as micro-cellular and pico-cellular require a dynamic range up to 90 and 55 dB respectively [6–9]. Thus making a highly-linear optical RoF transmitter necessary in order to achieve the required signal dynamic range [10]. Linearization techniques are often employed to the power amplifier transmitter to improve system performance.

Increasing demand for spectral efficiency in radio communications makes multi-level linear modulation schemes, such as quadrature amplitude modulation (QAM) and orthogonal frequency division multiplexing (OFDM), in more demand [3–5, 11]. However, the use of such linear modulation schemes have increased the linearity requirements of RF components such as power amplifiers. Since their signals' envelopes fluctuate, these modulation schemes are more sensitive to power amplifier nonlinearities, the major contributor of nonlinear distortion in a transmitter. Therefore the power amplifier is required to process high data rate non-constant envelope signals [11]. For achieving good power efficiency the power amplifier should work around its compression point which makes the output signal distorted nonlinearly. These nonlinear distortions generate in-band interferences which results in amplitude and phase deviation of the modulated vector signal. It also generates out-band interference in the adjacent channel creating spectrum

Z. El-Khatib et al., *Distributed CMOS Bidirectional Amplifiers: Broadbanding and Linearization Techniques*, Analog Circuits and Signal Processing, DOI 10.1007/978-1-4614-0272-5_1, © Springer Science+Business Media New York 2012

spreading. When a non-constant envelope signal goes through a nonlinear PA, spectral regrowth broadening appears in the PA output causing adjacent channel interference [11]. PA linearization is often necessary to suppress spectral regrowth and reduce bit error rate (BER).

Active MOSFET devices can be modeled by nonlinear current and charge sources that depend on the device voltages [12, 13]. These nonlinear sources will give rise to distortion when driven with a modulated signal. The real MOSFET device output impedance is nonlinear and the mobility μ is not a constant but a function of the vertical and horizontal electric field. We may bias the active MOSFET device where the device behavior is more exponential. And there is also an internal feedback so when the input signal driven into the amplifier is increased, the output is also increased until a point where distortion products can no longer be ignored [12, 14].

The harmonics of the output signal are generated by nonlinearities of the MOSFET devices. The major three nonlinear elements of the MOSFET devices are nonlinear transconductance g_m, the device drain capacitance C_d and gate capacitance C_{gs} [12, 15] as shown in Fig. 1.1c. The real MOSFET devices generate higher order distortion [12, 16]. Models are used to characterize the nonlinear behavior of a semiconductor device in order to predict the resultant signal properties. The simple polynomial approximation is a nonlinear transfer function based upon the Taylor series expansion [15, 16].

Ideal RF power amplifier has an output signal that is a precise scaled copy of the input signal. Unfortunately real amplifiers are characterized by some degree of nonlinearity. No transistor is perfectly linear since the inherent nonlinearity of the diode junctions that comprise many of the active devices found in most amplifiers. Equivalent circuit of n-section CMOS distributed amplifier including nonlinear generators in the drain and gate lines is shown in Fig. 1.1a. Fully-integrated broadband linearization techniques are required to address linearity improvements for UWB communication such as bidirectional RoF systems [17, 18]. Traditional broadband amplifiers only allow broadband amplification in one direction. However integrating the bidirectional element into a linearized DA allows for broadband amplification in both directions eliminating the need for RF switches that degrade performance and increase insertion loss. The versatility of broadband distributed amplifier (DA) configurations makes them useful for performing several circuit functions integrated in RoF systems and wireless transceivers as shown in Fig. 1.1b, c [19–22]. The data-carrying RF signal is imposed on the lightwave signal for modulating light before being transported over the optical link and then converted from optical to electrical domain at the base stations or remote antenna unit before being amplified and radiated by an antenna as shown in Fig. 1.1a. A mixer and oscillator components are added to the remote antenna unit for RF signal processing as shown in Fig. 1.1a. As can be depicted in Fig. 1.1b, an application of a CMOS bidirectional DA based active circulator has port(1) and port(4) as input ports and port(3) and port(4) as output ports.

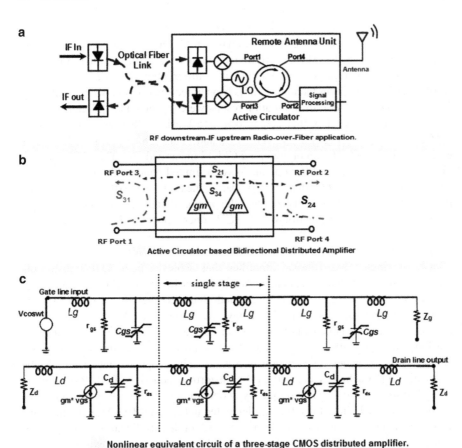

Fig. 1.1 (**a**) Application of CMOS distributed bidirectional amplifier integrated in bidirectional radio-over-fiber remote antenna unit with RF downstream and IF upstream transmission. (**b**) Active circulator based bidirectional distributed amplifier. (**c**) Equivalent circuit of n-section CMOS distributed amplifier including nonlinear generators in the drain and gate lines

In response to the need to correct the DA's nonlinear distortion, a number of DA linearization techniques have been developed and have been reported in literature [10–23]. However, most of the published DA linearization methods reported do not provide fully-integrated distortion cancellation technique with large third-order intermodulation (IM3) distortion reduction. Since they involve system-level linearization with bulky discrete components which is not suited for fully-integrated circuit miniaturization. Due to the discrete component performance variation with frequency, they suffer from limited linearization over broad bandwidth [10, 23]. Other DA linearization techniques involve circuit-level linearization, however they have narrow linearized bandwidth and apply DC-based linearization technique only [24, 25].

A CMOS DA based multi-tanh linearization technique is also reported in literature [26]. However the CMOS DA based multi-tanh linearization technique offered a limited 5 dB IM3 distortion reduction only [26]. Another linearized DA that has been reported in literature is a differential DA with circuit-level feedforward linearization technique [27]. It operated over a very wide band from 0.1 to 12 GHz, however only simulation results were presented [27]. Lau and Chan proposed a linearized DA that achieved a 10 dB IM3 reduction over a limited 2.3 GHz bandwidth range [23]. Recently, Lu and Pham [28, 29] proposed a multi-gated transistor (MGTR) topology based CMOS linearized DA. The MGTR-based linearized distributed amplifier operated over a limited bandwidth of 4 GHz range and had only a 11 dB IM3 reduction. Comparing the proposed CMOS linearized DA in this work [30] to other published ones, the proposed linearized CMOS DA offers a 20 dB IM3 distortion reduction with 9.5 GHz operational bandwidth and with the least power consumption [30].

In this book, we demonstrate a fully-integrated fully-differential linearized CMOS distributed bidirectional amplifier that achieves large IMD3 distortion reduction over broadband frequency range for both RF paths. The drain and gate transmission-lines were stagger-compensated. Reducing the DA IM3 distortion by mismatching the gate and drain LC delay-line ladders. A CMOS cross-coupled compensator transconductor is proposed, in Sect. 5.3, to enhance the linearity of the DA gain cell with a varactor-based active post nonlinear drain capacitance compensator for wider linearization bandwidth. The proposed linearized CMOS bidirectional DA achieves a measured IM3 distortion reduction of 20 dB over ultra-wideband from 0.1 to 9.5 GHz frequency of operation with a two-way amplification of 5 dB in both RF directions. It is implemented in 0.13 μm RF CMOS technology with a silicon chip area of 1.5 mm^2 for use in highly-linear UWB wireless communication systems and in bidirectional RoF transmission as shown in Fig. 1.1a.

1.2 Book Research Objectives

The main objective of this book research is aimed at the broadband amplifier linearization providing broadband linear amplification for use in UWB wireless communication systems and bidirectional RoF transmission. The specific research design goals in pursuit of the main objective are as follows:

1. The development of linearized CMOS stagger-compensated bidirectional distributed amplifier with 20 dB IM3 distortion reduction.
2. The development of a highly-linear CMOS cross-coupled compensator transconductor with enhanced tunability.
3. The application of linearization techniques to distributed circuit designs such as power splitters, matrix amplifier and paraphase amplifier.

1.3   Outline of the Book

The research objectives are introduced in Chap. 1. In Chap. 2, modulation schemes effects on RF power amplifier nonlinearity and RFPA linearization techniques are presented. In Chap. 3, distributed amplification principles and transconductor nonlinearity compensation are presented. Various applications of linearized distributed circuit functions are presented in Chap. 4. Chapter 5 describes in detail the proposed fully-integrated linearized CMOS bidirectional distributed amplifier and the proposed highly-linear CMOS cross-coupled compensator transconductor with enhanced tunability. Chapter 6 presents the proposed linearized CMOS bidirectional distributed amplifier layout techniques and considerations. Chapter 7 presents the proposed linearized CMOS bidirectional DA experimental test setups and measured results. Chapter 8 draws conclusions of the book work and lists research contributions.

Chapter 2
Modulation Schemes Effect on RF Power Amplifier Nonlinearity and RFPA Linearization Techniques

2.1 Introduction

In this chapter, RF modulation schemes effects on RF power amplifier nonlinearities are presented. A review of various power amplifier RF linearization techniques are discussed.

2.2 RF Modulation Scheme in Bandpass Radio Communication Channel

In radio communications, modulation can be described as the process of conveying a message signal by superimposing an information bearing signal onto a carrier signal by varying the signal characteristic [31–33]. Modulation is the process of changing a higher frequency signal in proportion to a lower frequency one or vise versa. The higher frequency signal is referred to as the carrier signal and the lower frequency signal is referred to as the information bearing message signal or modulating signal [31–33]. The characteristics (amplitude, frequency or phase) of the carrier signal are varied in accordance with the information bearing signal. These high-frequency carrier signals can be transmitted over the air, over fiber or coax cable. The use of high frequency signals will make the amplifier and antenna design easier for effective radio design [32, 33].

Figure 2.1 shows the up conversion of the complex-valued baseband signal x'(t) to the passband then the transmission of the real-valued bandpass signal x(t) through the communication channel [31]. After the bandpass signal goes through the channel, a down-conversion of the bandpass output Y(t) into a complex-valued baseband signal Y'(t) occurs. The baseband signal x'(t) is up-converted to the bandpass signal by amplitude, phase or frequency modulation in order to transmit it. The modulated bandpass signal x(t) can be described as

Z. El-Khatib et al., *Distributed CMOS Bidirectional Amplifiers: Broadbanding and Linearization Techniques*, Analog Circuits and Signal Processing, DOI 10.1007/978-1-4614-0272-5_2, © Springer Science+Business Media New York 2012

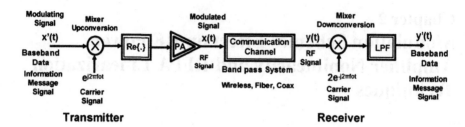

Fig. 2.1 A general universal illustration of a bandpass communication channel system

$$x(t) = A(t)\cos(2\pi f_o t + \phi(t)) \tag{2.1}$$

where f_o is the carrier frequency, A(t) is the amplitude and $\phi(t)$ is the phase modulation.

The bandpass signal x(t) has an envelope bandwidth lower then the carrier frequency f_o. Using trigonometric identities, this signal can be re-written as

$$x(t) = I(t)\cos(2\pi f_c t) - Q(t)\sin(2\pi f_c t) \tag{2.2}$$

where I(t) is the in-phase component and Q(t) is the quadrature component

$$I(t) = A(t)\cos[\theta(t)] \tag{2.3}$$

$$Q(t) = A(t)\sin[\theta(t)] \tag{2.4}$$

Consequently, the bandpass signal x(t) can be re-written in complex form as

$$x(t) = A(t)\cos[\theta(t)]\cos(2\pi f_c t) - A(t)\sin[\theta(t)]\sin(2\pi f_c t) \tag{2.5}$$

$$x(t) = A(t)\cos[2\pi f_c t + \theta(t)] = \text{Re}\left[A(t)\,e^{j\theta(t)}e^{j2\pi f_c t}\right] \tag{2.6}$$

$$x(t) = \text{Re}\left[x'(t)\,e^{j2\pi f_c t}\right] \tag{2.7}$$

where x'(t) is the baseband input signal and can be represented as

$$x'(t) = I(t) + jQ(t) \tag{2.8}$$

$$x'(t) = A(t)\,e^{j\theta(t)} \tag{2.9}$$

Therefore, the real-valued bandpass input signal x(t) can be obtained from the complex valued baseband input signal x'(t) as [31, 33]

$$x(t) = \text{Re}\left\{x'(t)\,e^{j2\pi f_o t}\right\} \tag{2.10}$$

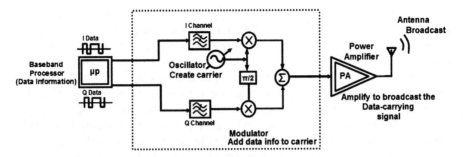

Fig. 2.2 Ideal radio quadrature upconverter transmitter

where f_o is the carrier frequency. Similarly, the baseband output signal y'(t) can be obtained from the bandpass output signal y(t) through demodulation process $2e^{-j2\pi f_o t}$.

The communication channel can be linear or nonlinear. The choice of a modulation scheme depends on the physical characteristics of this channel, required levels of performance and hardware trade-offs [33].

2.2.1 Ideal Radio Transmitter

As shown in Fig. 2.2 the baseband processor converts the information we want to send into data. The data can be either analog (continuously varying) or digital (in discrete states) in format [31, 33, 34].

An ideal radio quadrature upconverter transmitter generates a high-frequency carrier for the data information to ride on. To do this we use a component called an oscillator as shown in Fig. 2.2. Ideal radio quadrature upconverter transmitter as shown in Fig. 2.2 has an oscillator that converts DC bias into a radio-frequency carrier. Then the carrier is combined with the data using a component called a modulator [32–34]. The data adjusts or modulates the characteristics of the carrier (amplitude, frequency or phase) in a controlled manner. The third step is to increase the signal strength with a power amplifier so that it can be detected by the receiver. The output of the power amplifier feeds an antenna which broadcasts the information carrying signal into the air. It also can be transmitted through other communication mediums such as fiber or coax cable [32–34].

2.2.2 RF Power Amplifier Linearity for Non-Modulated Signal

Active MOSFET devices can be modeled by nonlinear current and charge sources that depend on the device voltages [12, 13]. These nonlinear sources will give rise to distortion when driven with a modulated signal. The real MOSFET device output impedance is nonlinear and the mobility μ is not a constant but a function of the vertical and horizontal electric field. We may bias the active MOSFET device where the device behavior is more exponential. And there is also an internal feedback so when the input signal driven into the amplifier is increased, the output is also increased until a point where distortion products can no longer be ignored [12, 14].

No transistor is perfectly linear since the inherent nonlinearity of the diode junctions that comprise many of the active devices found in most amplifiers. The harmonics of the output signal are generated by nonlinearities of the MOSFET devices. The major three nonlinear elements of the MOSFET devices are nonlinear transconductance g_m, the device drain capacitance C_d and gate capacitance C_{gs} [12, 15]. The real MOSFET devices generate higher order distortion [12, 16]. Models are used to characterize the nonlinear behavior of a semiconductor device in order to predict the resultant signal properties [15, 16].

Power amplifiers can be classified in to two categories, linear and nonlinear. Linear power amplifiers preserve amplitude and phase information where as nonlinear power amplifiers only preserve phase information [32, 33]. Linear power amplifiers employ transistors as current sources with high impedance. Nonlinear power amplifiers employ transistors as switches with low impedance. Linear power amplifiers can drive both broadband and narrowband loads. Nonlinear power amplifiers usually drive a tuned circuit narrowband load. Models are used to characterize the nonlinear behavior of a semiconductor device in order to predict the resultant signal properties. The simple polynomial approximation is a nonlinear transfer function based upon the Taylor series expansion. Typically the first-order (gain), second-order (squaring) and third-order (cubing) terms are considered [15, 16].

Power amplifier device nonlinearity can be modeled by a polynomial [14, 35]

$$v_o(t) = f(v_i(t)) = a_1 v_i(t) + a_2 v_i^2(t) + a_3 v_i^3(t) + \cdots a_N v_i^N(t) = \sum_{n=1}^{N} a_N v_i^N(t)$$
(2.11)

Applying a single-tone RF signal to the power amplifier transistor

$$v_i(t) = A_1 \cos(\omega_o t + \phi_1)$$
(2.12)

$$v_o(t) = a_1 A_1 \cos(\omega_o t + \phi_1) + a_2 A_1^2 \cos^2(\omega_o t + \phi_1) + \cdots + a_n A_1^n \cos^n(\omega_o t + \phi_1)$$
(2.13)

Writing out the response $v_o(t)$ by performing trigonometric expansion [36, 37]

$$v_o(t) = a_1 A_1 \cos(\omega_o t + \phi_1) + a_2 \frac{A_1^2}{2} - a_2 \frac{A_1^2}{2} \cos^2(2\omega_o t + 2\phi_1)$$

$$+ a_3 \frac{A_1^3}{4} \cos^2(3\omega_o t + 3\phi_1) + a_3 \frac{3A_1^3}{4} \cos^2(\omega_o t + \phi_1)$$
(2.14)

Table 2.1 One tone signal generated harmonics

$a_1 A_1 \cos(\omega_o t + \phi_1)$	Linear gain
$a_2 \frac{A_1^2}{2}$	DC offset (self-bias)
$a_2 \frac{A_1^2}{2} \cos^2(2\omega_o t + 2\phi_1)$	Second harmonic distortion
$a_3 \frac{A_1^3}{4} \cos^2(3\omega_o t + 3\phi_1)$	Third harmonic distortion
$a_3 \frac{3A_1^3}{4} \cos^2(\omega_o t + \phi_1)$	(AM AM and AM PM)

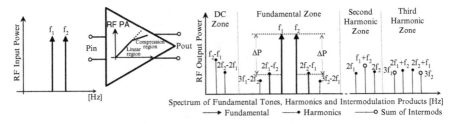

Fig. 2.3 Output spectrum of a power amplifier that includes the desired fundamental signals as well as the spurious products created by intermodulation distortion

Another method of testing power amplifier linearity is the two-tone method with two closely spaced fundamental signals tones applied to the test amplifier (Table 2.1). The amplitude is increased until the third-order cross-product produces a signal above the noise floor [32–34].

As shown in Fig. 2.3 the result of applying two tones to amplifiers that exhibit a degree of nonlinearity is intermodulation distortion (IMD) and third-order harmonics grouped in harmonic zones. As can be depicted from Fig. 2.4 that the third order intermodulation products $(2f_1 - f_2)$ and $(2f_2 - f_1)$ are the main contributor to distortion in that they are very near the fundamental tones and are not filtered out as is the case of the second order intermodulation products $(f_1 - f_2, 2f_1, f_1 + f_2$ and $2f_2)$ [32–34].

Typically the third-order intermodulation distortion product (IM3) are of most concern since distortion products which are far away in frequency from the desired output can be removed by filtering as shown in Fig. 2.4 [32–34].

Applying a two-tone RF signal to the power amplifier

$$v_i(t) = v \cos(\omega_1 t) + v \cos(\omega_2 t) \tag{2.15}$$

The two-tone signal covers the complete dynamic range of the amplifier (Table 2.2). The amplifier output is a power series expansion up to the fifth-order is given by Rogers and Plett [32]

$$v_o = a_1 v_i + a_2 v_i^2 + a_3 v_i^3 + a_4 v_i^4 + a_5 v_i^5 \cdots \tag{2.16}$$

The output voltage with two-tone signal

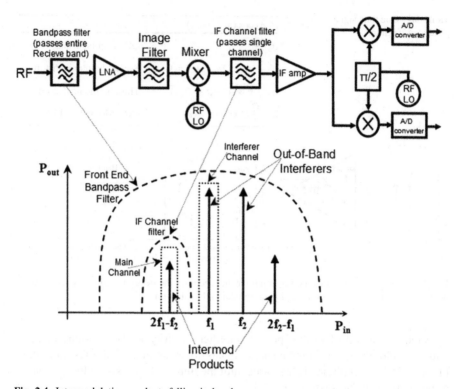

Fig. 2.4 Intermodulation products falling in-band

$$v_o(t) = a_1 v \left[\cos(\omega_1 t) + \cos(\omega_2 t)\right] + a_2 v^2 \left[\cos(\omega_1 t) + \cos(\omega_2 t)\right]^2$$
$$+ a_3 v^3 \left[\cos(\omega_1 t) + \cos(\omega_2 t)\right]^3 + a_4 v^4 \left[\cos(\omega_1 t) + \cos(\omega_2 t)\right]^4$$
$$+ a_5 v^5 \left[\cos(\omega_1 t) + \cos(\omega_2 t)\right]^5 \tag{2.17}$$

Figure 2.5 shows a plot of IM3 versus P_{in} and P_{out} versus P_{in}. If the 1:1 slope line of the fundamental Pout and the 1:3 slope line of the third order intermodulation product are extended, they will intersect at a point called IP3, the third-order intercept point. IP3 is an approximation because the slope assumption is not truly valid outside the linear region. The higher the IP3 point the less distortion at higher power levels. In the linear part of Fig. 2.5, the Pout versus Pin curve has a slope of 1:1. The P1dB point which is the power where the gain drops by 1 dB compared to the linear gain was another way to characterize power amplifiers [32–34].

In the nonlinear region in Fig. 2.5 higher efficiency is gained over 50%, however distortion is increased significantly. Constant amplitude modulation schemes like AMPS or FM radio use saturated power amplifiers that are more efficient than linear ones [32–34].

Table 2.2 Two tone signal generated harmonics coefficients

	a_1v	a_2v^2	a_3v^3	a_4v^4	a_5v^5
DC		1		9/4	
ω_1	1		9/4		25/4
ω_2	1		9/4		25/4
$2\omega_1$		1/2		2	
$2\omega_2$		1/2		2	
$\omega_1 \pm \omega_2$		1		3	
$2\omega_1 \pm \omega_2$			3/4		25/8
$\omega_1 \pm 2\omega_2$			3/4		25/8
$3\omega_1$			1/4		25/16
$3\omega_2$			1/4		25/16
$2\omega_1 \pm 2\omega_2$				3/4	
$\omega_1 \pm 3\omega_2$				1/2	
$3\omega_1 \pm \omega_2$				1/2	

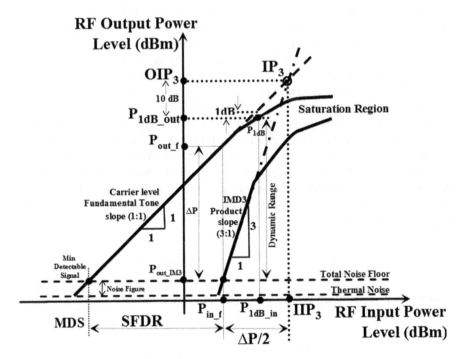

Fig. 2.5 Output power versus input power defining third-order intercept point IP3

$$P_{IM3} = 3 \times P_{IN} - 2 \times IIP3 \ (dBm) \tag{2.18}$$

$$OIP_3 = \frac{3 \times P_{out\_f} - P_{out\_IM3}}{2} \tag{2.19}$$

Dynamic range is the signal range in which the signal can still be processed with high quality. It is the signal range whose lower limit is defined by the sensitivity level and whose upper limit is defined by the acceptable maximum level of signal distortion [32–34]. SFDR (spurious free dynamic range) can be found from the two linear equations (2.18) and (2.19) for the harmonic and third-order intermodulation product where MDS is the minimum detectable signal [33].

$$SFDR = \frac{2}{3}(IIP3 - MDS) \tag{2.20}$$

All amplifiers have maximum output power capacity which is called saturation output power. There are a number of ways to specify the nonlinear behavior of a power amplifier. One method of defining amplifier's linearity is third-order intercept point (IP3) [32–34]. This method relies on a figure of merit that is determined by graphical extrapolation of amplifier data taken well below saturation. The IP3 is a theoretical point obtained by extending the two functions until they intersect. If an amplifier was operated at a given level below this third order intercept point then its linear performance was considered adequate. Figure 2.5 is a plot of the output versus input transfer function of the power amplifier whose desired fundamental outputs (f_1 and f_2) describe a function with a slope of one. The third order intermodulation products are also plotted in Fig. 2.5. The output level continues to increase with an increase of input power until a point is reached where output device begins to saturate resulting in a gradual roll-off of output power.

When the actual output power level differs by 1 dB compared to the ideal output value, the P1dB compression point is reached. In other words, P1dB is the output power level point at which the gain is 1 dB compressed. The P1dB is another main power amplifier identification parameter. The higher the intercept point, the better the amplifier is at amplifying large signals [32–34]. However, with variety of new modulation methods, the P1dB compression point is not enough for power amplifier performance prediction [32–34].

IIP3 can be described by

$$IP3_{|dBm} = P_{IN|dBm} + \frac{\Delta P_{|dB}}{2} \tag{2.21}$$

IP5 may be determined using the two-tone test as well. Similar to IP3 equation, IP5 can be described by

$$IP5_{|dBm} = P_{IN|dBm} + \frac{\Delta P_{|dB}}{4} \tag{2.22}$$

2.2.3 RF Power Amplifier Linearity for Modulated Signals

The new generation of mobile communication technologies employ linear modulation (QPSK, QAM) and wide bandwidth for increasing bit rate and spectrum efficiency. Therefore the power amplifier is required to process high data rate

Table 2.3 Different PAR values for different modulated signal types [1, 2]

Modulated signal type	Modulation method	PAR [dB]
CDMA	QPSK	10
TDMA	Π/4-DQPSK	3.5
2-tone	AM	3
1-tone	FM	0

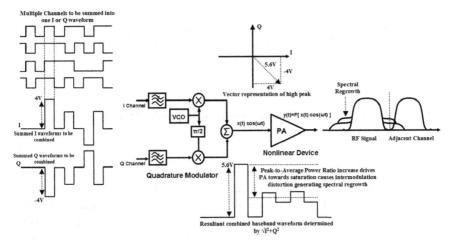

Fig. 2.6 RF power amplifier spectral regrowth

non-constant envelope signals. For achieving good power efficiency, the power amplifier should work around its compression point however making the output signal distorted nonlinearly [38, 39].

Spectrally efficient modulation schemes in wireless systems non-constant envelope signals have high peak to average power ratio. These modulation techniques require a highly linear PA to process high data rate non-constant envelope signals. In general the higher the data rates the higher the peak to average power ratio (PAR) [2]. High peaks can cause the power amplifier to move toward saturation. This causes intermodulation distortion which generates spectral regrowth as shown in Fig. 2.6. Spectral regrowth is a condition that interferes with signals in adjacent frequency bands and can be reduced by using power amplifier linearization techniques [38, 39]. Table 2.3 shows the different PAR values for different modulated signal types [1]. The PAR is a strong function of the type of modulation. For 1 dB of PAR that means operating the power amplifier at 1 dB lower power or power back-off which is not as power efficient [40].

Power peaks develop in wireless digital communication signal such as CDMA waveforms [2, 38, 39]. A wireless digital communication signal is generated from a quadrature modulator as shown in Fig. 2.6. The waveform is composed of an I waveform and a Q waveform and these waveforms are the summation of multiple channels. Whenever the channel waveforms simultaneously contain a bit in the same state a high power peak occurs in the summed waveform as shown in Fig. 2.6.

The I and Q waveforms combine in the quadrature modulator to create an RF waveform [38,39]. The magnitude of the RF envelope is determined by the I squared plus Q squared modulator equation where the squaring of I and Q always results in a positive value. The simultaneous positive peaks in the I and Q waveforms combine to create a greater peak as shown in Fig. 2.6.

Power amplifier nonlinearity effects over modulated signals are out-of-band distortion and in-band distortion effects. The out-of-band distortion effects produces a spectrum widening which results in higher ACPR (Adjacent Channel Power Ratio) value. The in-band distortion effects produces a constellation distortion which results in higher BER (Bit Error Rate).

2.2.4 RF Power Amplifier Spectral Regrowth: Out-of-band Distortion

Adjacent-channel power ratio is the linearity figure-of merit for wireless communication systems employing non-constant envelope modulation techniques such as QAM and Π/4-DQPSK [38, 39]. These linear modulation techniques, although spectrally efficient, produce modulated carriers with envelope fluctuations. This envelope fluctuation results in signal distortion and spectral spreading when the modulated carrier is passed through a saturated RF power amplifier.

RF power amplifier nonlinear effect impacts the CDMA signal's out of band emission levels. A general mathematical model of a CDMA signal's spectrum, s(t) can be described as [38,39,41,42]

$$s(t) = \sqrt{2}x(t)\cos(2\pi f_o t + \theta) \tag{2.23}$$

where x(t) is a base-band white Gaussian process with phase θ.

The input–output relationship can be approximated using Taylor polynomial of the input signal for a weak nonlinearity. A general mathematical model of a RF power amplifier can be described as [41,43]

$$y(t) = F(s(t)) = a_1 s(t) + a_3 s^3(t) \tag{2.24}$$

A nonlinear power amplifier transfer function leads to odd-order intermodulation products. These third-order intermodulation products cause distortion. Only the odd-order terms in the Taylor series are considered. The effect of spectra generated by the even-order terms on the passband are negligible since they are at least f_c away from the center of the passband.

Coefficient a_1 describes the linear gain of the amplifier and a_3 is the nonlinear coefficient. For a linear power amplifier the expression for the a_1 and a_3 coefficients can be expressed as [41,43]

$$a_1 = 10^{\frac{G}{20}} \tag{2.25}$$

and

$$a_3 = \frac{2}{3} 10^{\left(\frac{-IP3}{10} + 3\frac{G}{20}\right)} \tag{2.26}$$

Coefficient a_3 is by far the major contributor to the distortion in a power amplifier since it's product appear around carrier frequency f_c.

The spectrum $Py(f)$ of the RF power amplifier can be expressed as [41, 43]

$$Py(f) = \frac{1}{2B}\left[Po - 6P_o^2 10^{-\frac{IP3}{10}} + 9P_o^3 10^{-\frac{IP3}{5}}\right]$$

$$+ \frac{3}{4B^3} P_o^3 10^{-\frac{IP3}{5}} \left[6B^2 - (f - f_c)^2\right], |f - f_c| \leq B \tag{2.27}$$

$$Py(f) = \frac{3}{8B^3} P_o^3 10^{-\frac{IP3}{5}} (3B - |f - f_c|)^2, B < |f - f_c| \leq 3B \tag{2.28}$$

$$Py(f) = 0 \tag{2.29}$$

when

$$3B < |f - f_c| \leq 3B \tag{2.30}$$

Using the results from $Py(f)$, the emission power level within the band P_{IM3} can be described as [43, 44]

$$P_{IM3} = \int_{f_1}^{f_2} Py(f) \, df = \frac{1}{8B^3} P_o^3 10^{-\frac{IP3}{5}} \left[(3B - f_1)^3 - (3B - f_2)^3\right] \tag{2.31}$$

IP3 can be expressed in terms of P_{IM3} as [38, 39, 44]

$$IP3 = -5\log\left[\frac{P_{IM3}(f_1 - f_2)B^3}{P_{IM3}\left[(3B - f_1)^3 - (3B - f_2)^3\right]}\right] + 22.2 dBm \tag{2.32}$$

Equation 2.32 describes IP3 in terms of the out-of-band emission power of a CDMA signal power amplifier [43]. Figure 2.7 Effects of PA nonlinearity on adjacent channel power ratio. Figure 2.8 shows the predicted output power spectral density (PSD) spectrum for CDMA signal of an RF power amplifier and its distortion effects. A nonlinear power amplifier transfer function leads to odd-order intermodulation products in which these third-order intermodulation products cause spreading of the distortion. These effects lead to spectral regrowth in adjacent channels [38, 39, 44].

Adjacent Channel Power Ratio (ACPR) is defined as the ratio of the main channel output power to the power in the adjacent channel [38, 42]. ACPR helps in determining the amount of signal energy leaked from the main channel to the adjacent channel and is used in testing of CDMA-based communication systems [45].

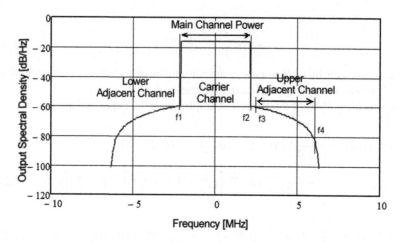

Fig. 2.7 Effects of PA nonlinearity on adjacent channel power ratio

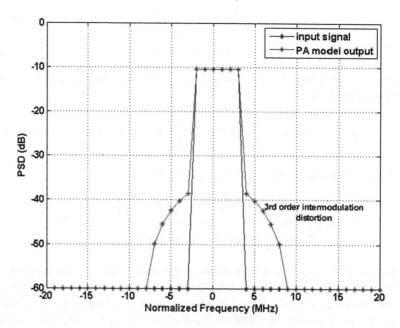

Fig. 2.8 Effects of PA nonlinearity on adjacent channel power ratio with third-order distortion

$$ACPR = \frac{\int\limits_{adjacent\_channel} P_{out}(f).df}{\int\limits_{main\_channel} P_{out}(f).df} \qquad (2.33)$$

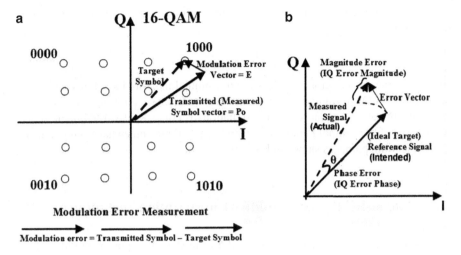

Fig. 2.9 Effects of PA nonlinearity on error vector magnitude

High crest factor can be defined as [38, 39, 44]

$$\xi = 10\log\left(\frac{P_{peak}}{P_{average}}\right) \tag{2.34}$$

$$ACPR = -20.75dB + 1.6\xi + 2\left(P_{in} - IP3\right) \tag{2.35}$$

RF power amplifier linearization techniques can enhance the overall system response of non-constant modulated signals by reducing ACPR in the PA output power spectral density.

2.2.5   Error Vector Magnitude Signal Modulation Quality: In-band Distortion

For achieving good power efficiency, the power amplifier should work around its compression point which makes the output signal distorted nonlinearly. These nonlinear distortions generate in-band interferences which results in amplitude and phase deviation of the modulated vector signal. In band interference causes errors in the symbol vectors. While ACPR describes the effects of nonlinearity on other channels, the error vector magnitude (EVM) is used to analyze in-band distortion [46]. EVM is the measure between the ideal reference target symbol vector and the transmitted measured symbol vector as shown in Fig. 2.9. Signal and error vectors are defined in the I-Q constellation diagram. EVM is defined as a percentage of peak signal level, it measures the modulation quality of the signal and indicates modulation accuracy.

The ratio of the error vector magnitude to the original symbol magnitude defines the EVM as

$$EVM = \frac{E}{P_o} \tag{2.36}$$

where E is the error vector and P_o is the transmitted measured symbol vector. An unimpaired 16-QAM digitally modulated signal would have all of its symbols land at exactly the same 16 points on the constellation over time. Real-world impairments cause most of the symbol landing points to be spread out somewhat from the ideal symbol landing points as shown in Fig. 2.9a.

2.3 Role of RF Power Amplifier Linearization Techniques

This section will discuss techniques for the cancellation of power amplifier distortion that are also known as linearization. Linearization of power amplifiers for radios, using advanced modulation schemes with large peak-to-average ratios, is important in achieving high efficiency as shown in Fig. 2.10. Many linearization techniques have been developed to improve power amplifier linearity and to decrease adjacent channel interference ACPR [44]. The basic idea is to run the power amplifier as close to saturation as possible to maximize its power efficiency, and then employ some linearization technique to suppress the distortion introduced in this near saturated region [35].

There are many linearization techniques for minimizing power amplifier nonlinear distortion. Linearization can be conducted at either circuit-level or sub-system

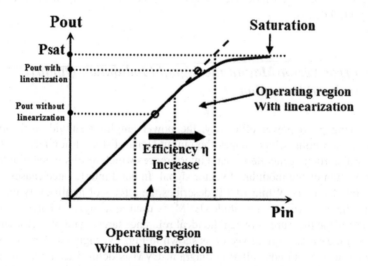

Fig. 2.10 Role of linearization techniques

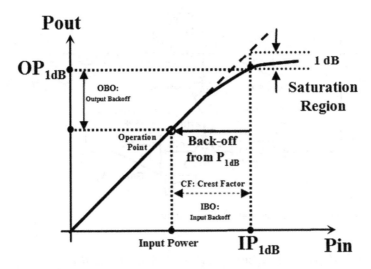

Fig. 2.11 RF power amplifier power back-off

level. Linearization techniques such as RF power amplifier power Back-off, RF predistortion, feedback and feedforward are often used. Here is a brief review of the major techniques for improving power amplifier linearity [35].

2.3.1   RF Power Amplifier Power Back-off

A simple way to improve power amplifier linearity performance is the Back-off operation [35]. To reduce distortion to an acceptable level one must operate the power amplifier at reduced power level (back-off from saturation). The back-off is the distance between the saturated point and the average power level. Increasing the back-off of the power amplifier means that the signal is contained better in the linear range, and thus the effects of nonlinearities are reduced. However, power efficiency is reduced as well. When a power amplifier is driven with decreased input power, the linearity of the power amplifier is improved as shown in Fig. 2.11.

$$Input\ Power = IP_{1dB} - IBO \qquad (2.37)$$

The benefits of the Back-off linearization technique is that it is simple. However, output back-off results in poor efficiency of DC-to-RF power conversion. Since efficiency has a high impact on cellular units talk time, allowances for output back-off have significantly been reduced [35]. So a trade off between efficiency and linearity must be made. Linearization techniques prove to be the best solution in order to improve power amplifiers linearity without having negative impact on efficiency as shown in Fig. 2.10.

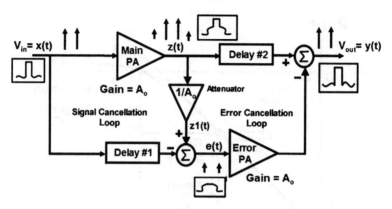

Fig. 2.12 RF power amplifier feedforward linearization

2.3.2 RF Power Amplifier Feedforward Linearization

The feedforward linearization technique was invented by H. S. Black and has since found applications in many communication systems [35] and [47]. The feedforward linearization architecture is shown in Fig. 2.12 and it is based on splitting the input signal x(t) into two branches. In the main branch the input signal x(t) is amplified by the main power amplifier yielding the PA output z(t). In the secondary branch the PA output z(t) is scaled and compared with the original input x(t) [35].

The resulting error signal e(t) goes through a second PA known as the error PA. After the error signal e(t) is obtained it is amplified and subtracted from the delayed output of the main PA. Since the error signal e(t) is the nonlinear distortion, removing it from the PA output linearizes the PA [35].

The following equations describes the feedforward linearization. Once the PA output $z(t)$ is attenuated to the same level of the input signal $x(t)$ by Webster and Parker [35]

$$z1(t) = {}^{z\,(t)}\!/_{A_0} \tag{2.38}$$

then obtaining the distortion by comparing $z1(t)$ with the input signal $x(t)$

$$e(t) = x(t) - z1(t) = x(t) - {}^{z(t)}\!/_{A_0} \tag{2.39}$$

The distortion can be amplified with the auxiliary error PA and subtracted from the original PA output as shown in Fig. 2.12

$$y(t) = z(t) + e(t) \cdot A_0 = z(t) + \left[x(t) - \frac{z(t)}{A_0} \right] \cdot A_0 = A_0 \cdot x(t) \tag{2.40}$$

Feedforward linearization is stable however it suffers from poor efficiency since an auxiliary error PA is needed.

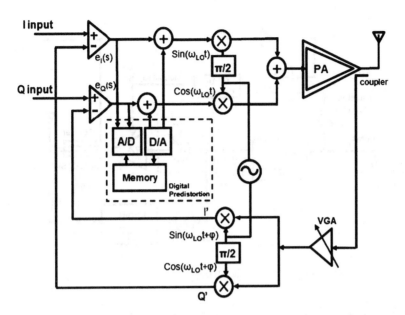

Fig. 2.13 RF power amplifier digitally assisted cartesian feedback linearization

2.3.3 RF Power Amplifier Cartesian Indirect Feedback Linearization

Many types of feedback linearization techniques exist including cartesian feedback and polar feedback [7, 35, 48]. Cartesian feedback linearization technique is based on feedback control system [7].

In cartesian feedback linearization as shown in Fig. 2.13 [2], the distorted PA output is fed back through an I-Q demodulator to build two negative feedback loops [49]. The distorted RF signal from the antenna is split into distorted Cartesian components I' and Q' [35]. The undistorted I and Q signal from the input and the distorted I' and Q' signal from the antenna are fed into differential amplifiers. The differential amplifiers compare two input signals and the amplified error signal e(s) are up-converted to an RF signal using a modulator. Because the feedback technique takes the output of the power amplifier as a reference during the correction process and it overcomes the behavior variation of the power amplifier [35]. The main objective of the cartesian feedback is to keep feedback phase φ aligned with the input signal phase. Cartesian feedback technique is simple and power efficient however it suffers from limited bandwidth [2, 35, 50].

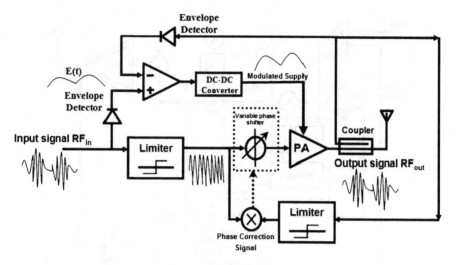

Fig. 2.14 RF power amplifier polar feedback linearization

2.3.4 RF Power Amplifier Polar Feedback Linearization

Like cartesian feedback the polar feedback linearization technique is a baseband feedback scheme [47, 51]. The polar feedback uses the magnitude and the phase of the PA output signal as feedback signals opposed to the cartesian feedback which uses inphase and quadrature signals [35, 52]. It seems more appropriate to use the envelope signal as one of the feedback signals because the distortion (AM/AM and AM/PM) is directly related to this signal [35, 47, 51, 52].

First the output from the PA is down converted to an intermediate frequency as shown in Fig. 2.14. Then the envelope of the signal is extracted. This is carried out by mixing the signal with an amplitude limited version of itself [35]. The amplitude of the input signal and the PA signal is subtracted to calculate the amplitude error. This signal is fed into the power regulator of the PA where the amplitude is adjusted. The phase error is calculated by the phase detector and used to adjust the phase of the VCO. This way the phase distortion of the PA is compensated by a phase change in the VCO. Polar feedback linearization technique provides relatively high efficiency since the power amplifier can operate completely nonlinearly [35].

2.3.5 RF Power Amplifier RF Predistortion Linearization

RF Predistortion is a popular linearization technique [53], it actively tracks and applies an inverse to the amplifier nonlinearity [11]. If the amplifier exhibits a gain compression the predistortion linearizer is designed to have a gain expansion

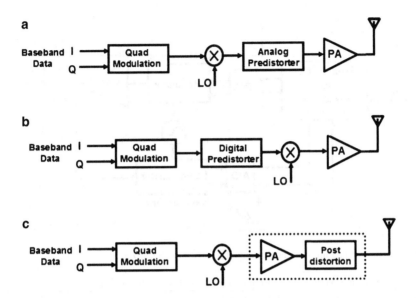

Fig. 2.15 RF power amplifier RF predistortion linearization

characteristic. The linearizer can be either active, passive, shunt or series [35]. As input power is increasing it will absorb less power to compensate the power gain roll-off of the following PA. When the input power is decreasing the effect is reversed. The fundamental principle of these predistortion techniques is to adjust the amount of input power [6].

An RF predistortion system uses an active or passive analog nonlinear element operating at the radio carrier frequency to generate the predistorted signal as shown in Fig. 2.15. IF predistortion implements the predistortion at some intermediate frequency allowing the system to be used at a number of different carrier frequencies [18]. Another implementation of predistortion is baseband predistortion where the inverse transfer function is applied prior to upconversion of the signal.

A possible simplified implementation of adaptive digital predistortion is shown in Fig. 2.16 [2]. Adaptive digital predistortion solves the problem of power amplifier variations in RF predistortion. It maintains a dynamically updated model of the power amplifier. Adaptive digital predistortion has an advantage of not suffering from bandwidth limitations incurred by feedback techniques [35, 54]. A comparison of different RF power amplifier linearization techniques is shown in Table 2.4 [2, 35, 52].

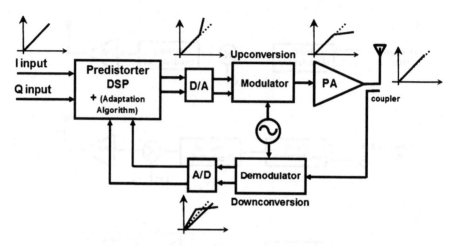

Fig. 2.16 RF power amplifier adaptive digital predistortion linearization

Table 2.4 Comparison of different RF power amplifier linearization techniques

Linearization technique	Linearization performance	Compensation bandwidth	Cost	Issues	Control applied at
Feedforward	Good	Wider	High	Low efficiency	Output
Cartesian indirect feedback	Moderate	Narrow	Moderate	Reduced gain stability	Input
Analog RF predistortion	Low	Wide	Low	Reduced gain	Input
Digital predistortion	Moderate	Wide	Moderate	Easy to control depends on DSP	Input
Polar feedback	Moderate	Wide	Moderate	Reduced gain	Input

2.4 Radio-over-Fiber for Wireless Communication

Radio-over-Fibre (RoF) is a technology that integrates radio and optics [4,5]. Radio transmission over fiber is used in cable television networks and in satellite base stations. The RoF system generally consists of commonly used components such as light sources, modulators, photo-detectors and optical fibers [3,6,7].

In RoF architecture, light is modulated by a data-carrying radio signal and transmitted over an optical fiber link as shown in Fig. 2.17. RF signals are optically distributed to base stations directly at high frequencies and converted to electrical domain at the base stations before being amplified and radiated by an antenna [3–8]. In IF-over-Fiber architecture, an IF (Intermediate Frequency) radio signal with a lower frequency is used for modulating light before being transported over the optical link as shown in Fig. 2.18.

Some of the advantages of RoF system are low attenuation, lower cost and low complexity [3, 4, 6, 7, 55]. In the basic RoF transmitter design, light intensity is modulated by the (modulated) RF signal, either directly modulating the laser current

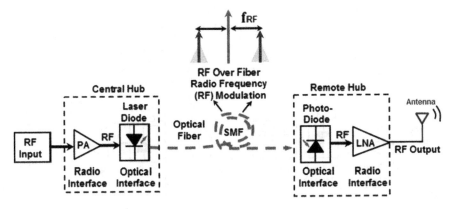

Fig. 2.17 Direct modulation of laser with RF over fiber radio frequency RF modulation

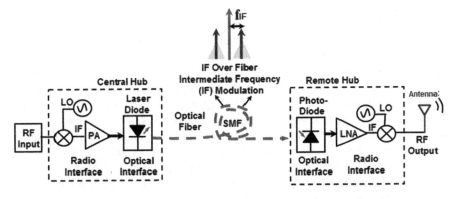

Fig. 2.18 Direct modulation of laser with if over fiber intermediate frequency modulation

or by applying an external modulator. Typically direct modulation is available for frequencies less than 1 GHz. External modulators such as the MachZehnder modulator are used for higher frequencies such as 1 GHz as shown in Fig. 2.19 but at additional cost [3, 5–7]. At the receiver side, the intensity of the transmitted RF signal guided by the optical fibre is detected with the aid of a photodetector and such a process is commonly referred to as intensity-modulation direct-detection (IM/DD) [6–8].

In RoF the optical fiber is used to carry RF signals. Since RoF involves detection of light, it is fundamentally an analog transmission system. Although the RoF transmission system itself is analog, the radio system being distributed does not have to be analog as well. It may be digital using comprehensive multi-level signal modulation formats such as QAM or Orthogonal Frequency Division Multiplexing (OFDM). RF communication systems use advanced forms of modulation to increase the amount of data that can be transmitted in a given amount of frequency spectrum by combining multiple single-carrier into a single transmitter such as OFDM [3, 5].

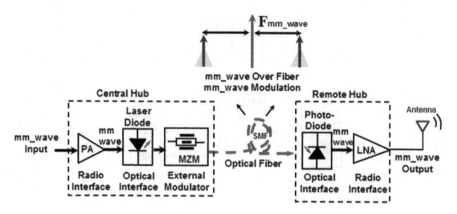

Fig. 2.19 External modulators are used for mm-wave over fiber modulation

Radio technologies such as ultra wideband (UWB) are able to provide high bit-rates. UWB RoF utilize OFDM modulation scheme for high rate networks with data rates reaching up to 480-Mbps [3, 4, 6, 7, 55].

However, combining several carrier signals within the transmitter power amplifier creates large variations in the instantaneous output power, a condition described as high peak-to-average ratio (PAR) [3, 5]. Signal impairments such as distortion, which are important in analogue communication systems, are important in RoF systems as well. Large PAR can generate performance degradation and transmitter power amplifier may exhibit nonlinear behavior as a result of the high PAR. These impairments tend to limit the dynamic range of the RoF links. Dynamic range is an important parameter for wireless communication systems because the power received at the base station varies widely [3–5, 55].

In applications requiring a linear PA due to PAR, back-off from the peak power point must be applied to avoid clipping the waveform. Another way to minimize distortion is to apply power amplifier linearization techniques.

2.5 Summary

In this chapter, RF modulation schemes effect on RF power amplifier nonlinearities was presented. RF power amplifier linearity for non-modulated and modulated signals was presented as well. RF power amplifier spectral regrowth out-of-band distortion and in-band distortion was discussed. A review of various power amplifier RF linearization techniques such as feedforward, cartesian feedback, polar feedback and RF predistortion were discussed.

Chapter 3
Distributed Amplification Principles and Transconductor Nonlinearity Compensation

3.1 Introduction

In this chapter, analysis of the distributed amplification principles are presented in this chapter. Different linearized transconductors to compensate for the nonlinearity are also presented.

3.2 Distributed Amplification Principles

A distributed amplifier is recognized as one of the most popular broadband amplifier designs. A detailed paper by Ginzton et al. [56] turned Percival's patented concept of distributed amplification into an actual implementation. A high gain-bandwidth product is usually the aim in amplifier design. The gain-bandwidth product is proportional to transconductance over capacitance and is defined as [16, 46, 57–59]

$$G \cdot Bw = \frac{g_m}{C} \qquad (3.1)$$

3.2.1 Additive Distributed Versus Product Cascaded Amplification

In a cascaded amplifier system, if the amplifier stages are identical, the overall bandwidth of N stages f_n is related to single stage 3 dB bandwidth f_i by Refs. [16, 57–59]

$$f_n = 0.83 \times \frac{f_i}{\sqrt{N}} \qquad (3.2)$$

where N is the number of amplifier stages and f_n the overall bandwidth of N stages.

Z. El-Khatib et al., *Distributed CMOS Bidirectional Amplifiers: Broadbanding and Linearization Techniques*, Analog Circuits and Signal Processing, DOI 10.1007/978-1-4614-0272-5_3, © Springer Science+Business Media New York 2012

The gain-bandwidth product GBP of a single stage in a cascaded amplifier system is given by Refs. [16, 57–59]

$$GBP = \frac{g_m}{2\pi C_D} \tag{3.3}$$

where g_m is the amplifier transconductance.

The gain A_i of a single stage and the GBP in a cascaded amplifier system is given by Refs. [16, 57–59]

$$A_i \times f_i = GBP \tag{3.4}$$

The gain A_n of N stage cascaded amplifier system is given by Refs. [16, 57–59]

$$A_n = (A_i)^N = \left(\frac{GBP}{f_i}\right)^N = \left(\frac{0.83 \times GBP}{\sqrt{N}\, f_n}\right)^N \tag{3.5}$$

When the number of stages N is increased in a cascaded amplifier system the gain increases however the bandwidth decreases.

For a distributed amplifier the gain is calculated as follows [46, 57, 60, 61]

$$A = \frac{N g_m Z_{od}}{2} \tag{3.6}$$

where Z_{od} is given by

$$Z_{od} = \frac{1}{2\pi C_d f_c} \tag{3.7}$$

For the distributed amplifier the gain is proportional to the number of stages and the bandwidth increases as well. Distributed amplification allows operation beyond cut-off frequency [57, 60–62]. The gain-bandwidth product of a distributed amplifier is limited by f_{max} and can be shown to be [57] Fig. 3.2.

$$A \cdot f_{1dB} = 0.8 f_{max} \tag{3.8}$$

where f_{1dB} is the frequency at which the gain falls 1 dB below nominal bandwidth. The maximum available power gain cut-off frequency f_{max} is given by Wong [57]

$$f_{max} = \frac{f_T}{2\sqrt{\frac{R_i}{R_{ds}}}} \tag{3.9}$$

where f_T is the unity current gain cut-off frequency and it is given by

$$f_T = \frac{g_m}{2\pi C_{gs}} \tag{3.10}$$

where g_m is the device transconductance and the C_{gs} is the device gate capacitance.

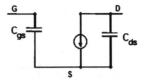

MOS devices have capacitive input and output components.

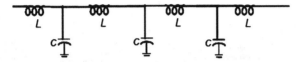

A low-pass transmission line constructed of inductors and capacitors

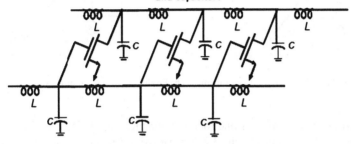

Input and output capacitances of the amplifier can be used to replace the transmission line capacitors

Fig. 3.1 A transmission-line constructed of inductors and capacitors coupled with amplification MOS devices

Combining amplifiers in parallel does not help as it increases the total capacitance. MOS devices have capacitive input and output impedances as depicted in Fig. 3.1. However, these capacitances can be incorporated in or counted as capacitors in a transmission-line. A physical structure that guides an electromagnetic wave from place to place is called transmission line [57,60]. A low-pass transmission-line can easily be constructed of inductors and capacitors and two transmission-lines can be coupled by amplifiers as can be seen in Fig. 3.1. Input and output capacitances of the amplifier can be used to replace the transmission-line capacitors [57,60,61].

Conventional distributed amplifiers use low-pass π-sections to form an artificial transmission line topology (Fig. 3.1). Amplification gain stages are connected so that output currents are combined in an additive manner at the output terminal [57, 60, 61]. The advantages of a distributed amplifier topology are its wide bandwidth, flat gain and compact size. It Provides a good isolation from output to input resulting in a stable amplifier configuration with no oscillation tendency. It also Provides a good input and output match so gain stages can be cascaded. The disadvantages of the distributed amplifier topology are its higher power consumption and lower efficiency.

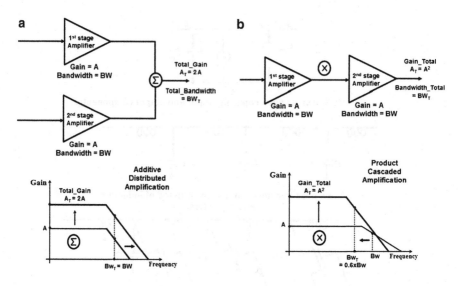

Fig. 3.2 (**a**) Additive distributed amplification. (**b**) Product cascaded amplification

The artificial transmission-line topology enables distributed amplifiers to have wider bandwidth [57, 60, 61] Fig. 3.2. Transmission-line capacitors, in conjunction with inductors coupled by the transconductance of the FETs provide a propagating medium in which signal waves can travel [63]. If these artificial transmission-lines are well designed, the maximum operating frequency of the amplifier would be limited by the cut-off frequency of the transmission-line [57, 60, 61]

$$f_c = \frac{1}{\pi\sqrt{LC}} \tag{3.11}$$

where L is the transmission-line lumped series inductance and C is the lumped shunt capacitance per section.

The transconductance networks isolate the shunt capacitance of the transistors from one another. The capacitances form the integral parts of a filter structure, whose bandwidth is determined by the amount of inductance and capacitance of the filter section. The gain can be increased by introducing more and more sections. We can now trade delay, rather than bandwidth, for gain. As the RF input signal travels down the gate transmission-line, each FET transistor is excited by the traveling power wave and transfers the signals to the drain line through its transconductance [57, 60, 61].

If the phase velocities on the gate and drain lines are equal, then the power signals on the drain line add in the forward direction and an RF output signal is generated [61, 64–66]. The waves traveling in the reverse direction are absorbed by the drain line termination [21, 57, 60, 61]. Artificial transmission-lines are usually modeled by cascading T-sections as shown Fig. 3.3a. This structure is usually referred to

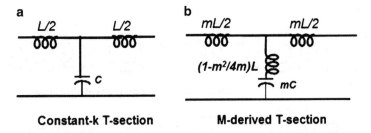

Fig. 3.3 (a) Constant-k T-sections (b) m-derived T-sections

as low-pass Constant-k T-section filter. Constant-k T-section transmission-lines are matched to a load using an m-derived section in order to provide constant Z-impedance over a wider range as can be seen in Fig. 3.3b. The parameter m is usually equal to 0.6 and is identified as a practical rule of thumb value. An m-derived section with parameter m equal to 1 corresponds to a constant-k T-section [57, 60, 61].

The image impedance Z_i, for a reciprocal symmetric two-port, can be defined as the impedance looking into port 1 or 2 of a general two-port network when the other terminal is also terminated in Z_i [67]. To achieve an impedance match over a broad range, the load and source impedance must be transformed into the image impedance Z_i. The m-derived section serves this purpose well. The m-derived impedance matching network provides an improvement to the variation over the broadband frequency. It can also be used to match directly to $Z_o = 50 \, \Omega$. The impedance looking into the gate and drain line when transformed by the m-derived section is approximately constant over a broad range of frequencies [57, 60, 61].

3.2.2 Lumped Constant Delay Line Characteristics

Transmission lines that are designed to intentionally introduce a time delay in the path of an electromagnetic wave are known as delay line. Delay lines are made of lumped elements of L and C. The inductors are connected in series and the capacitors are connected from the junctions between inductors to the ground. The parasitic L and the C from the transistors are used as well. The lumped constant delay line can be considered as a special purpose low pass filter composed of series inductors and shunt capacitors used to delay (phase shift) the input signal by a specified increment of time (degrees). Time delay can be realized with lumped LC delay-lines. The delay of the line, T_d, is a function of the total inductance and capacitance [57, 60, 61].

The LC artificial transmission line time delay of each section as shown Fig. 3.4 can be expressed as inductance and capacitance per section [46, 57, 60, 61, 67–69]

$$T_d = \sqrt{LC} \tag{3.12}$$

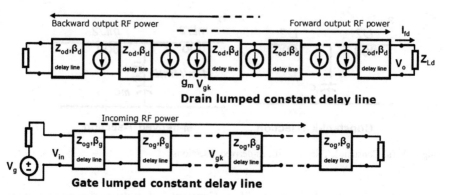

Fig. 3.4 Lumped constant delay line characteristics

where L is the transmission-line lumped inductance and C is the lumped capacitance per section.

The input and output delay lines are designed to achieve a characteristic impedance of [46, 57, 60, 68, 69]

$$Zo = \sqrt{L/C} \tag{3.13}$$

The phase velocity of the LC delay lines can be expressed as [46, 57, 60, 68, 69]

$$v_p = \frac{1}{\sqrt{LC}} \tag{3.14}$$

The LC delay line also forms a low-pass filter whose cutoff frequency is [46, 57, 60, 68, 69]

$$f_c = \frac{1}{\pi \sqrt{LC}} \tag{3.15}$$

The propagation constant of the LC delay lines is in general a complex quantity and so may be defined as [46, 57, 60, 68, 69]

$$\gamma = \alpha + j\beta \tag{3.16}$$

α is called the attenuation constant and is measured in decibels or nepers per unit length of the transmission line. The propagation constant γ is a measure of the phase shift and attenuation per unit length along the line. β known as the phase constant is the phase shift per unit length of transmission line and is measured in radians per unit length of this line [46, 57, 60, 68, 69].

$$\beta = 2\pi/\lambda \tag{3.17}$$

where λ is the distance along the line corresponding to a phase change of 2π radians.

The phase constant β is expressed as [46, 57, 60, 68, 69]

$$\beta = \omega\sqrt{LC} \tag{3.18}$$

The propagation velocity on the line or phase velocity of a wave is the rate at which the phase of the wave propagates in space. The phase velocity is given in terms of the wavelength (lambda) and period T or inverse of time delay [46, 57, 60, 68, 69]

$$v_p = \frac{\lambda}{T} = \frac{\omega}{\beta} = \frac{1}{\sqrt{LC}} \tag{3.19}$$

The input voltage can be described as [46, 57, 60, 68, 69]

$$v(t) = V\cos(\omega t) \tag{3.20}$$

where ω is the angular frequency (rad/sec).

The traveling wave can be described as [46, 57, 60, 68, 69]

$$v(z, t) = V\cos(\omega t - \beta z) \tag{3.21}$$

where β is the propagation constant (rad/m).

If the line is terminated in characteristic impedance, the traveling wave voltage distribution on the line will be reduced due to attenuation [46, 57, 60, 68, 69].

$$v(z, t) = Ve^{-\alpha z}\cos(\omega t - \beta z) \tag{3.22}$$

where α is the attenuation coefficient (nepers/meter).

A phasor can be used to represent the amplitude of a sinusoidal voltage. This phasor does not include any frequency representation [46, 57, 60, 68, 69].

$$v(z, t) = Ve^{-\alpha z}\cos(\omega t - \beta z) = \text{Re}\left(Ve^{-\alpha z}e^{-j\beta z}e^{j\omega t}\right) \tag{3.23}$$

The voltage wave traveling down the lumped constant delay line, as shown Fig. 3.4, is continuously attenuated if the line is terminated in Z_o the characteristic impedance of the line. The voltage at the nth section is given by Refs. [46, 57, 60, 68, 69]

$$V_o = V_{in}e^{-\gamma n} \tag{3.24}$$

where V_{in} is a sinusoidally varying input function.

$$V_o = V_{in}e^{-\alpha n}e^{-j\beta n} \tag{3.25}$$

The delays of the input and output lines can be made equal through the selection of propagation constants. Delay mismatches occur when the output currents do not

add in phase due to the phase delay in the gate being different than the one in the
drain [46,57,60,68,69].

The gain for a distributed amplifier with N transistors assuming zero attenuation
is given by Refs. [46,57,60,68,69]

$$\frac{V_o}{V_{in}} = \frac{g_m Z_o}{2} \sum_{n=1}^{N} e^{-j(n-1/2)(\beta_g - \beta_d)} \tag{3.26}$$

It can be seen that the gain becomes maximum when $\beta_g - \beta_d = 0$

3.2.3   Lossless Distributed Amplification

In the case of a lossless lumped cascade of T-section, the image impedance and the
propagation coefficient are respectively given by Refs. [46,57,60,68,69]

$$Z_i = \sqrt{L/C \left(1 - \left(\omega/\omega_c\right)^2\right)} \tag{3.27}$$

$$e^\gamma = 1 - \frac{2\omega^2}{\omega_c^2} + \frac{2\omega}{\omega_c^2}\sqrt{\frac{\omega^2}{\omega_c^2} - 1} \tag{3.28}$$

where ω_c is the line cut-off frequency of the transmission line given by Refs. [46,
57,60,68,69]

$$\omega_c = \frac{2}{\sqrt{LC}}. \tag{3.29}$$

The characteristic impedance Z_o of the lossless artificial transmission-line is given
by Refs. [46,57,60,68,69]

$$Z_o = \sqrt{\frac{L}{C}} \tag{3.30}$$

where L and C are the inductance and capacitance per unit length of the
transmission-line. The artificial transmission-line has a cut-off frequency f_c given
by Refs. [46,57,60,68,69]

$$f_c = \frac{1}{\pi\sqrt{LC}}. \tag{3.31}$$

The bandwidth is determined by the input capacitance of the transistor and the
inductance of the transmission-line yielding a wider bandwidth than that of lumped
element circuits. The C_{gs} and C_{ds} of MOS device capacitances are absorbed into
the transmission-line.

To get a better understanding of the CMOS DA operation, it useful to consider
the wave propagation behavior along the lossless transmission line. Assuming ideal

transistors with no parasitics, the common-source amplifier output drain current on the drain line is given by Refs. [16, 64, 66]

$$i_d = g_m \cdot v_{gs} \tag{3.32}$$

The RF input signal is fed into the gate line, where it excites the FET transistors along the gate line as it propagates, until it reaches the gate line termination, where it is ideally dissipated without reflections [42, 70].

Assuming gate and drain line have equal propagation velocity and zero loss, $\beta_g = \beta_d$ and $\alpha = 0$ for the lossless condition, where α is defined as attenuation factor and β is also known as phase constant [13, 39, 61, 64, 70].

An input wave v_g propagating on a lossless line can be expressed as in (3.33) [39, 46], where v_g appears at the transistor inputs as v_{gs} and half of V_s being the amplitude of the input wave that flows into the terminating gate resistance [61].

$$v_g(z) = \frac{V_s}{2} \cdot e^{-j\beta z} \tag{3.33}$$

An input wave propagating along the gate line to the output on the drain line is given by Refs. [60, 61]

$$v_{gn} = \frac{V_s}{2} \cdot e^{-(n-1)j\beta_g \ell_g} \tag{3.34}$$

where $n = 1$ when it is the first stage, implies there is no delay. The output drain current and output voltage relationship is given in (3.35) [60, 61]

$$v_o = I_d \cdot Z_d \tag{3.35}$$

The output drain current across the n-th FET, I_d, can be expressed as [46, 47, 61]

$$I_d = \frac{1}{2} \sum_{n=1}^{N} i_{dn} \left(e^{-(N-n)} \right)^{j\beta_d \ell_d} \tag{3.36}$$

where $n = N$ represents the last stage [60, 61]

$$i_{dn} = g_m \cdot v_{gs} \tag{3.37}$$

Substituting (3.34) and (3.36) into (3.37) we get [46, 47, 61]

$$I_d = -\frac{g_m}{4} \cdot V_s \cdot \sum_{n=1}^{N} \left(e^{-(n-1)} \right)^{j\beta_g \ell_g} \cdot \left(e^{-(N-n)} \right)^{j\beta_d \ell_d} \tag{3.38}$$

$$I_d = -\frac{g_m}{4} \cdot V_s \cdot e^{-Nj\beta_d \ell_d} \cdot e^{j\beta_g \ell_g} \cdot \sum_{n=1}^{N} \left(e^{-nj} \right)^{(\beta_g \ell_g - \beta_d \ell_d)} \tag{3.39}$$

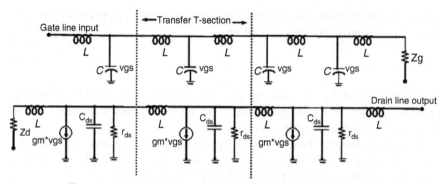

T-sections structure of gate and drain lines of the distributed amplifier.

Fig. 3.5 T-sections structure of gate and drain lines of the distributed amplifier

Each FET injects current into the output line as seen in Fig. 3.5, if the phase constant of both transmission-lines are equal, the currents add constructively at the output, while out-of-phase signals are dissipated on the drain termination [46]. If under normal operating conditions, the waves in the gate and drain lines are synchronized [71]

$$\beta_g \ell_g = \beta_d \ell_d = \beta \ell \tag{3.40}$$

$$I_d = -\frac{g_m}{4} \cdot V_s \cdot e^{-Nj\beta\ell} \cdot e^{j\beta\ell} \cdot N \tag{3.41}$$

In the case of ideal lossless constant-k lines, the expression for the power gain of N section distributed amplifier is thus given by Heutmaker and Michael [46]

$$G = \frac{P_{out}}{P_{in}} = \frac{\frac{1}{2}|I_d|^2 Z_d}{|v_g|^2 / 4Z_g} = \frac{g_m^2 Z_d Z_g N^2}{4} \tag{3.42}$$

For the ideal lossless transmission line case the gain increases relative to N^2 as can be seen from (3.42), and the bandwidth is infinite.

3.2.4 Lossy Distributed Amplification

In the presence of attenuation, the gate signal decays as it propagates down the line. Hence, there will be a point at which the gain added by an additional device will not overcome the losses induced by the extra section in the gate and drain lines. The reason for this is that the devices added are not driven sufficiently to overcome the losses in the drain line cause of high signal attenuation.

A transmission-line can be constructed by a cascaded T-section structure with a characteristic impedance of [57, 60–62]

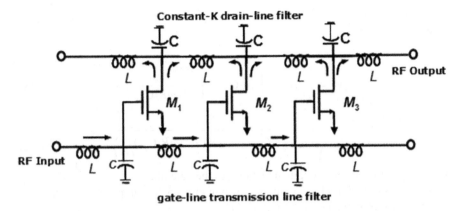

Fig. 3.6 Gate and drain transmission lines of a CMOS distributed amplifier

$$Z_{OT} = \sqrt{\frac{L}{C}} \frac{1}{\left(1 - \left(\frac{\omega}{\omega_c}\right)^2\right)^{\frac{3}{2}}} \sqrt{\left(1 - \left(\frac{\omega}{\omega_r}\right)^2\right)^2 - \left(\frac{\omega}{\omega_c}\right)^2} \qquad (3.43)$$

where $\omega_c = \frac{2}{\sqrt{LC}}$ is the cut-off frequency and $\omega_r = \frac{1}{\sqrt{LC_d}}$ is the drain line inductor self-resonance frequency. The presence of gate resistance r_g and drain resistance r_{ds} affects the frequency variation of the transmission-line characteristic impedance Z_{OT}.

Figure 3.5 shows an equivalent gate line circuit.

A common-source amplifier gate resistance R_i in series with parasitic capacitance C_{gs} introduces a frequency dependent attenuation per line section [46]. The impedance of the gate line, shown in Fig. 3.6, can be approximated as [57,60–62,71]

$$Z = j\omega L_g \qquad (3.44)$$

where L_g is the per-unit-length inductance of the gate line as shown in Fig. 3.5.

The propagation constant of the gate line, γ_g can be expressed as [13]

$$\gamma_g = \alpha_g + j\beta_g = \sqrt{Z \cdot Y} \qquad (3.45)$$

where the gate transmission-line equivalent impedance Z and admittance Y parameters are given by Refs. [13,57]

$$Y = j w C g + \frac{j w C_{gs}/l_g}{1 + j w R_i C_{gs}} \qquad (3.46)$$

$$Z_g = \sqrt{\frac{Z}{Y}} \cong \sqrt{\frac{L_g}{C_g + C_{gs}/l_g}} \qquad (3.47)$$

Substituting (3.46) and (3.47) into (3.45) we get the propagation constant of the gate line as follows

$$\gamma_g = \sqrt{j\omega L_g \left(j\omega C_g + \frac{j\omega C_{gs}/l_g}{1 + j\omega R_i C_{gs}} \right)} \tag{3.48}$$

$$\gamma_g \cong \sqrt{-\omega^2 L_g \left(C_g + C_{gs} \left(1 - j\omega R_i C_{gs} \right)/l_g \right)} \tag{3.49}$$

where L_g is the per-unit-length inductance of the gate line.

The equivalent drain line circuit is shown in Fig. 3.5. The impedance of the drain line can be approximated as [57, 60–62, 71]

$$Z = j\omega L_d \tag{3.50}$$

where L_d is the per-unit-length inductance of the drain line.

The propagation constant of the drain line, γ_d can be expressed as [13]

$$\gamma_d = \alpha_d + j\beta_d = \sqrt{Z \cdot Y} \tag{3.51}$$

where the drain transmission-line equivalent impedance Z and admittance Y parameters are given by Refs. [13, 57]

$$Y = \frac{1}{R_{ds} l_d} + j\omega \left(C_d + C_{ds}/l_d \right) \tag{3.52}$$

where R_{ds} is the shunt resistance of the MOSFET and ω is the operating frequency [57, 60–62, 71]

$$Z_d = \sqrt{\frac{Z}{Y}} \cong \sqrt{\frac{L_d}{C_d + C_{ds}/l_d}} \tag{3.53}$$

Substituting (3.52) and (3.53) into (3.51) we get the propagation constant of the drain line as follows

$$\gamma_d = \sqrt{j\omega L_d \left(\frac{1}{R_{ds} l_d} + j\omega \left(C_d + C_{ds} \right)/l_d \right)} \tag{3.54}$$

Assuming the $j\omega(C_d + C_{ds}/l_d)$ term dominates in (3.54), the propagation constant of the drain line, γ_d can be expressed as

$$\gamma_d \cong \frac{Z_d}{2 R_{ds} l_d} + j\omega \sqrt{l_d \left(C_d + C_{ds}/l_d \right)} \tag{3.55}$$

The output drain current with the lossy condition can be approximated as follows [57,60–62,71]

$$I_d = -\frac{g_m}{4} \cdot V_s \cdot e^{-N\gamma_d \ell_d} \cdot e^{\gamma_g \ell_g} \cdot \sum_{n=1}^{N} (e^{-n})^{(\gamma_g \ell_g - \gamma_d \ell_d)} \tag{3.56}$$

$$I_d = -\frac{g_m V_s}{4} \cdot \frac{e^{-N\gamma_g \ell_g} - e^{-N\gamma_d \ell_d}}{e^{-\gamma_g \ell_g} - e^{-\gamma_d \ell_d}} \tag{3.57}$$

When losses are incorporated in the analysis, the power gain can be determined as, assuming unilateral devices [57,60–62,71]

$$G = \frac{g_m^2 \cdot Z_d \cdot Z_g}{4} \cdot \frac{\left(e^{-N\alpha_g \ell_g} - e^{-N\alpha_d \ell_d}\right)^2}{\left(e^{-\alpha_g \ell_g} - e^{-\alpha_d \ell_d}\right)^2} \tag{3.58}$$

where α_g and α_d are the attenuation per section in Nepers on the gate and drain line.

$$e^{-\alpha_g \ell_g} - e^{-\alpha_d \ell_d} \approx \alpha_g \ell_g - \alpha_d \ell_d \tag{3.59}$$

Equation 3.58 indicates that the gain is no longer a monotonic function of N, the number of DA gain stage sections, but peaks at a finite N. In other words, adding an extra stage will only improve gain if the product of $\frac{g_m \cdot V_{gs}(N+1) \cdot Z_d}{2}$ is greater than $\left|e^{\gamma_d \ell_d}\right|$, where $\frac{g_m \cdot V_{gs}(N+1) \cdot Z_d}{2}$ is the extra gain and $e^{\gamma_d \ell_d}$ is the attenuation on the drain line. Therefore, the loss due to the extra transmission-line exceeds the gain of the Nth FET. Eventually the polynomial additive increase in DA gain due to the extra stage will not outperform the exponential loss.

By taking the derivative of the gain function in (3.58), we can then find the optimum number of CMOS gain cells for maximum gain [57,60–62,71]

$$N_{opt} = \frac{\ln\left(\alpha_g \ell_g / \alpha_d \ell_d\right)}{\alpha_g \ell_g - \alpha_d \ell_d} \tag{3.60}$$

where α_g and α_d are the attenuation per section in Nepers on the gate and drain line. We also notice that the gain is not flat with frequency due to frequency dependence of α_g and α_d.

An increase in number of sections, n, increases gain linearly. However, line losses and parasitics prevent an infinite increase in DA gain stages. The attenuation on the gate line will drive the input signal to negligibly small values and adding further stages will increase the attenuation on the drain line.

The expression for the forward power gain of a DA is given by Refs. [57,60–62,71]

$$G_{for} = \frac{g_m^2 Z_{og} Z_{od} \sinh^2 \left[\frac{n}{2} (\alpha_d - \alpha_g)\right] e^{-n(\alpha_d + \alpha_g)/2}}{4 \left[1 + \left(\varpi/\varpi_g\right)^2\right]^{1/2} \left[1 - \left(\varpi/\varpi_g\right)^2\right] \sinh^2 \left[\frac{1}{2} (\alpha_d - \alpha_g)\right]} \tag{3.61}$$

where Z_{og} and Z_{od} are the characteristic impedance of the gate and drain lines, respectively, α_g and β_g are the attenuation and phase shift per section on the gate line, α_d and β_d are the attenuation and phase shift per section on the gate line and ϖ_g is the gate cut-off frequency. In order to maximize the forward gain and minimize the ripple, α_g would be set equal to α_d. As the number of sections increases, the signal level on the gate line is attenuated and so is the power to the drain line. The loss in the drain line includes an exponential factor that is a function of the per section drain line attenuation.

The reverse gain in a DA can be written as [19, 20]

$$G_{rev} = \frac{g_m^2 Z_{og} Z_{od} e^{-n(\alpha_d + \alpha_g)} \left(\cosh\left(n (\alpha_d - \alpha_g)\right) - \cos 2n\beta\right)}{4 \left[1 + \left(\varpi/\varpi_g\right)^2\right]^{1/2} \left[1 - \left(\varpi/\varpi_c\right)^2\right] \left(\cosh (\alpha_d + \alpha_g) - \cos 2\beta\right)}. \tag{3.62}$$

The DA isolation is dependent on G_{rev}. The directivity can be derived as [19, 20]

$$D = 10 \log \frac{G_{for}}{G_{rev}} = 10 \log \left[\left(\frac{\sin \frac{n}{2} (\theta_d - \theta_g)}{\sin \frac{1}{2} (\theta_d - \theta_g)}\right)^2 \left(\frac{\sin \frac{1}{2} (\theta_d + \theta_g)}{\sin \frac{n}{2} (\theta_d + \theta_g)}\right)^2\right] \tag{3.63}$$

3.3 Transconductor Gain Cells for Fully-Differential Distributed Amplifiers

Active MOSFET devices can be modeled by nonlinear current and charge sources that depend on the device voltages [12, 13]. These nonlinear sources will give rise to distortion when driven with a modulated signal. The real MOSFET device output impedance is nonlinear and the mobility μ is not a constant but a function of the vertical and horizontal electric field. We may bias the active MOSFET device where the device behavior is more exponential. And there is also an internal feedback so when the input signal driven into the amplifier is increased, the output is also increased until a point where distortion products can no longer be ignored [12, 14].

No transistor is perfectly linear since the inherent nonlinearity of the diode junctions that comprise many of the active devices found in most amplifiers. The harmonics of the output signal are generated by nonlinearities of the MOSFET devices. The major three nonlinear elements of the MOSFET devices are nonlinear transconductance g_m, the device drain capacitance C_d and gate capacitance C_{gs}

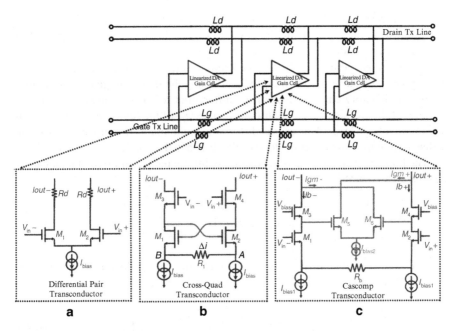

Fig. 3.7 (a) CMOS source degeneration transconductor. (b) Caprio's cross-quad transconductor. (c) Quinn's cascomp transconductor as a gain cell application for high-frequency amplifiers linearity enhancement

[12, 15]. The real MOSFET devices generate higher order distortion [12, 16]. Models are used to characterize the nonlinear behavior of a semiconductor device in order to predict the resultant signal properties [15, 16].

External linearization circuitry can be added to the amplifier, to compensate for the nonlinearity [34, 72]. This allows the nonlinear amplifier to be used to amplify signals utilizing spectrally efficient linear modulation techniques without causing interference.

In radio transceivers, a differential pair transconductor is often used to convert the RF input voltage signal to a current that is either amplified or frequency translated. External linearization circuitry can be added to the differential pair amplifier in order to compensate for the transconductance nonlinearity. This allows the nonlinear amplifier to be used to amplify signals utilizing spectrally efficient linear modulation techniques without causing interference. Many techniques have been developed to improve the linearity of the basic differential pair transconductor such as source degeneration, Caprio's cross-quad [21] and Quinn's cascomp transconductor [73] as shown in Fig. 3.7a–c respectively. However, their transconductance second-derivative gm″ nulling is not sufficiently wide and their gm flatness is limited [63, 74]. Other published linearized BJT V-I converters, such as the proposed linearized transconductor in [75], does not linearize the total transconductor drain current instead only linearizing the inner translinear loop current. Other published linearized transconductors lack on-chip tuning [56].

A three-stage bidirectional distributed amplifier with a differential pair transconductor as a gain stage is shown Fig. 3.7a coupling the drain and gate transmission-lines. The linearity characteristic of the differential pair is determined by the non ideality of the transistors. The transconductance g_m of the MOS device varies with the input voltage level. This causes distortions in the drain current which results in a distorted output voltage at the load of the distributed amplifier.

The transfer characteristic function of the differential pair transconductor can be expressed as [16, 58]

$$\Delta V_{out} = -\frac{1}{2}\mu_n C_{ox}\frac{W}{L}R_d \Delta V_{in}\sqrt{\frac{4I_{bias}}{\mu_n C_{ox}\frac{W}{L}} - \Delta V_{in}^2} \tag{3.64}$$

The small-signal gain A_v of the differential pair transconductor can be obtained through differentiating the transfer characteristic function in (3.64)

$$A_v = \frac{\delta V_{out}}{\delta V_{in}} = -R_d\sqrt{\mu_n C_{ox}\frac{W}{L}I_{bias}} = -R_d G_m \tag{3.65}$$

where $G_m = g_{m1} = g_{m2}$ is the overall transconductance of the differential pair transconductor.

The linearity characteristic of the differential pair transconductor is determined by the non-ideality of the transistors. To observe the linearity characteristics of the differential amplifier transconductor, its transfer function in (3.64) can be modeled by a third order polynomial Taylor series

$$V_{out} = a_1 V_{in} + a_2 V_{in}^2 + a_3 V_{in}^3 \tag{3.66}$$

The polynomial coefficients can be simplified by partial derivatives

$$a_1 = \frac{\partial V_{out}}{\partial V_{in}}\Big|_{vin=0} = -Rd\sqrt{\mu_n C_{ox}\frac{W}{L}I_{bias}} \tag{3.67}$$

$$a_2 = \frac{\partial^2 V_{out}}{\partial V_{in}^2}\Big|_{vin=0} = 0 \tag{3.68}$$

$$a_3 = \frac{\partial^3 V_{out}}{\partial V_{in}^3}\Big|_{vin=0} = \frac{3}{4}R_d\frac{\left(\mu_n C_{ox}\frac{W}{L}\right)^{\frac{3}{2}}}{\sqrt{I_{bias}}} \tag{3.69}$$

Since the differential pair transconductor generates no second order nonlinearity. The denominator of a_3 in (3.69) consists of the square root of bias current I_{bias}. Consequently, increasing the bias current I_{bias} lowers the coefficient a_3 and improves the linearity of the differential pair transconductor and hence overall distributed amplifier linearity. However, improving the linearity by increasing bias current I_{bias} has its limits. A higher I_{bias} increase power consumption.

The overall linearity of the DA can be improved as well by using Caprio's cross-quad transconductor as a gain stage shown in Fig. 3.7b. Caprio proposed a precision differential voltage-current converter linearization technique [21]. The cross-quad linearization technique shown in Fig. 3.7b uses a unity gain positive feedback in cross quad to synthesize a virtual ground across a passive element R_1. The positive feedback Caprio's cross-quad linearization technique is independent of g_m as in conventional degeneration method. Assuming perfect matching and ignoring body effects, the gate-source voltages of M_3 and M_2 are compensated by gate-source voltages of M_4 and M_1 shown in Fig. 3.7b.

$$V_A = \left(V_{in}^+/2\right) - V_{GS2} - V_{GS3} \qquad (3.70)$$

where V_{GS2} is the gate-source voltage of transistor M_2 and V_{GS3} is the gate-source voltage of transistor M_3 of Caprio's transconductor shown in Fig. 3.7b.

$$V_B = \left(V_{in}^+/2\right) - V_{GS1} - V_{GS4} \qquad (3.71)$$

where V_{GS1} is the gate-source voltage of transistor M_1 and V_{GS4} is the gate-source voltage of transistor M_4 of Caprio's transconductor shown in Fig. 3.7b.

Therefore, the overall Caprio's transconductor drain current is linearly proportional to the input voltage

$$\Delta i = \left(\frac{V_A - V_B}{R_1}\right) = \frac{V_{in}}{R_1} \qquad (3.72)$$

where Δi is the overall Caprio's transconductor drain current and V_{in} is the input voltage.

In Caprio's cross-quad, all four transistors M_1–M_4 have the same (W/L) transistor size and are perfectly matched. The gate-source voltages for M_1–M_3 are the same as the gate-source voltages for M_2–M_4 since the current flowing through the left branch of the cross quad is the common to M_1–M_3, while the current flowing through the right branch is common to M_2–M_4.

The overall linearity of the DA can be improved by using the cascomp (cascode compensator) transconductor as shown in Fig. 3.7c. A three-stage bidirectional DA with a cascomp transconductor as a gain stage is shown in Fig. 3.7c. The Cascomp transconductor circuit as shown in Fig. 3.7c for feedforward linearization was proposed by Quinn [73]. Ignoring body effects, the inner differential pair M_5 and M_6 subtracts the difference in gate-source voltages of M_1 and M_2 to linearize the transconductance [73] as shown in Fig. 3.7c.

$$V_{Rb} = V_{in} + V_{GS1} - V_{GS2} = V_{in} - \Delta V_{GS} = V_{Rb}\Delta i \qquad (3.73)$$

where V_{GS1} is the gate-source voltage of transistor M_1 and V_{GS2} is the gate-source voltage of transistor M_2 of Cascomp transconductor shown in Fig. 3.7c.

$$V_{AB} = (V_{BIAS} - V_{GS3}) - (V_{BIAS} - V_{GS4}) = V_{GS1} - V_{GS2} \qquad (3.74)$$

where V_{GS3} is the gate-source voltage of transistor M_3 and V_{GS4} is the gate-source voltage of transistor M_4 of Cascomp transconductor shown in Fig. 3.7c.

Once ΔV_{GS} is applied to the differential pair of M_5 and M_6, it drives the inverse drain currents for M_3 and M_4. Therefore, if

$$g_{m5} = g_{m6} = 2/Rb \qquad (3.75)$$

$$i_1 - i_2 = 2\frac{V_{in}}{R_b} \qquad (3.76)$$

Therefore, the overall Cascomp transconductor drain current is linearly proportional to the input voltage.

3.4   Chapter Summary

In this chapter, analysis of the distributed amplification principles were presented in this chapter. Different linearized transconductors to compensate for the nonlinearity such as cascomp and caprio's transconductors were also presented.

Chapter 4
Distributed RF Linearization Circuit Applications

4.1 Introduction

The interest in high-level integration and highly linear multi-functional broadband subsystems motivates the development of linearized distributed circuit applications. In this chapter, the application of RF linearization to distributed circuit functions such as fully-integrated CMOS linearized distributed active power splitter (unbalanced input, balanced output) and a linearized CMOS distributed matrix amplifier are presented. Also the linearized CMOS distributed paraphase amplifier employing derivative superposition linearization is presented as well.

4.2 Linearized CMOS Distributed Active Power Splitter

The design of a fully-integrated CMOS distributed active power splitter (unbalanced input, balanced output) incorporating multiple-gated transistor linearization that allows for broadband distortion reduction is presented in this section. Broadband power splitters are common elements found in many communication systems. Power splitters are found in phased antenna arrays [76, 77]. The Wilkinson power splitter was invented around 1960 by an engineer named Ernest Wilkinson [78, 79]. It splits an input signal into two equal phase output signals however the passive Wilkinson power splitter commonly used has an inherent 3 dB loss. The active power splitter provide an attractive alternative to the conventional Wilkinson or other planar power splitters in terms of gain, isolation and size. Hence, am active ultra-wideband circuit that can perform the same splitting function is presented.

The CMOS linearized distributed power splitter provides two outputs, with equal phase over a wide microwave band with minimal phase and amplitude imbalance. Compared to passive power splitter circuits, the linearized CMOS distributed power splitter's advantage is that it provides gain and allows for broadband distortion

Z. El-Khatib et al., *Distributed CMOS Bidirectional Amplifiers: Broadbanding and Linearization Techniques*, Analog Circuits and Signal Processing,
DOI 10.1007/978-1-4614-0272-5_4, © Springer Science+Business Media New York 2012

cancellation. The proposed CMOS linearized distributed power splitter makes up for the inherent 3 dB loss of passive power splitters and be used to drive balanced antennas and wideband phase shifters.

The wideband distributed power splitter is based on the concept of the distributed amplifier [19, 20]. The distributed amplifier is a classical topology for wideband amplifiers since the distributed amplifier has several attractive characteristics such as flat gain, good input and output matching as well as small size. The distributed amplifier design guidelines given by Beyer [20] can be extended to include linearized CMOS distributed power splitter. Figure 4.1 shows a schematic of an n-stage distributed power splitter and its equivalent circuit. It consists of n-stage distributed FET amplifiers. The balanced output transmission lines both have a distributed common-source amplifier. They share a common unbalanced input transmission line. A simple circuit model for the distributed active power splitter is shown Fig. 4.1.

The power gain of the common-source section of the distributed power splitter shown in Fig. 4.1 is given by [20]

$$G_{CS} = \frac{G_m^2 \, R_{og} \, R_{od}}{4 \left[1 + (\varpi/\varpi_g)^2\right]^{1/2} \left[1 - (\varpi/\varpi_c)^2\right]}$$

$$\times \frac{\sinh^2\left[\frac{n}{2}(\alpha_d - \alpha_g)\right] e^{-n(\alpha_d + \alpha_g)}}{\sinh^2\left[\frac{1}{2}(\alpha_d - \alpha_g)\right]} \tag{4.1}$$

where G_m is the transconductance and α_d and α_g are the attenuation per section and R_{od} and R_{og} are the characteristic resistance and ϖ_d and ϖ_g are of the drain and gate line respectively.

The amplitude imbalance of the distributed active power splitter can be expressed as [20]

$$\Delta A \, (dB) = 20 \log \frac{1 + \frac{1}{G_m R_d}}{1 + \frac{Z_{d\pi}}{2R_d}} + 20 \log \frac{\cosh\left(\frac{n}{2}\frac{\alpha_d}{\alpha_o}\right)}{\cosh\left(\frac{1}{2}\frac{\alpha_d}{\alpha_o}\right)}$$

$$-10 \log \left[\exp\left(n\frac{\alpha_d}{\alpha_o}\right)\right] \tag{4.2}$$

α_d and α_g are the attenuation per section and G_m is the transconductance.

The first term results from the gain difference between the two stages while the latter two terms stem from the difference in attenuation on the two output lines.

The phase imbalance of more practical concern is given by [20]

$$\delta\phi = \arctan \frac{\varpi C_d}{\frac{1}{R_d} + G_m} - \arctan \frac{\varpi C_d}{\frac{1}{R_d} + \frac{2}{Z_{d\pi}}} \tag{4.3}$$

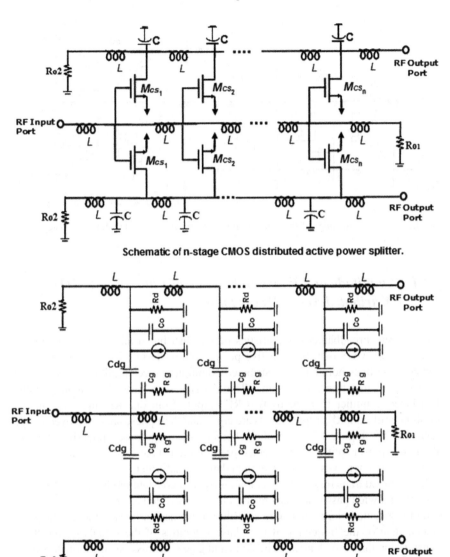

Schematic of n-stage CMOS distributed active power splitter.

Equivalent circuit of n-stage CMOS distributed active power splitter.

Fig. 4.1 Block diagram illustration of CMOS distributed active power splitter

where G_m is the transconductance. The critical design parameters are output line impedance level and device G_m transconductance. For instance, if output line impedance is dictated by circuit constraints the device can be chosen to minimize the phase imbalance.

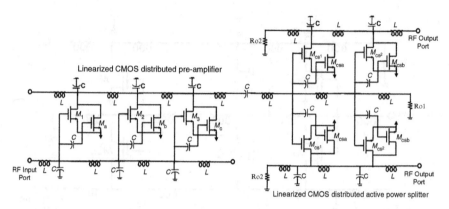

Fig. 4.2 Schematic topology of the proposed fully-integrated linearized CMOS distributed active power splitter

4.2.1 Amplitude and Phase Imbalance of Linearized CMOS Distributed Active Power Splitter

The inherent broadband characteristics of the distributed amplifier is applied to a linearized CMOS distributed active power splitter design. The balanced output transmission lines both have a distributed common-source amplifier. They are fed by a single input transmission line. The characteristic low pass response of the distributed amplifier then yields two RF outputs from one RF input signal, the amplitudes of both RF output signals are equal and their phase are equal over a wide band [78, 79]. Figure 4.2 shows the schematic of the proposed linearized CMOS distributed power splitter. A distributed pre-amplifier stage is added which not only supplies gain but serves as an active impedance transformer.

The lumped transmission lines tend to provide a constant input and output resistance over a wide passband. As a result, the phase difference between the balanced output distributed common-source amplifier transmission lines is theoretically 0°, independent of the frequency, if the phase velocities of the signals on the three transmission lines are identical. The drain and gate transmission-line inductors have an inductance value of 1.1 nH and the drain transmission-line m-derived inductance is 120 pH with 50 Ohms terminations. The device dimensions $[M_1, M_3]$ (W/L) and $[M_{cs1}, M_{cs2}]$ (W/L) have a width equal to 44 μm with all devices having L minimum channel length of 120 nm. The device dimensions $[M_a, M_c]$ (W/L) and $[M_{csa}, M_{csb}]$ (W/L) have a width equal to 22 μm with all devices having L minimum channel length of 120 nm. The power gains of the common-source amplifier on both outputs are almost the same with nearly equal output power as can be seen in Fig. 4.3 and as a result, a wideband CMOS distributed power splitter is obtained.

The linearized CMOS distributed power splitter has a 7.8 dB S_{21} power gain peak as shown in Fig. 4.3 and rolls off to a unity gain bandwidth of 10.5 GHz. The phase imbalance is less than 5° as shown in Fig. 4.4 and the amplitude imbalance

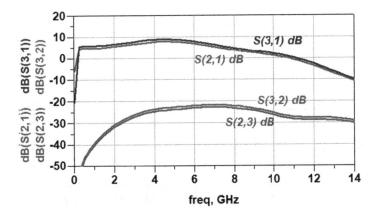

Fig. 4.3 Simulated output power S21 common-source stage port and S31 common-source stage port peaks at 7.8 dB for the proposed linearized CMOS distributed power splitter

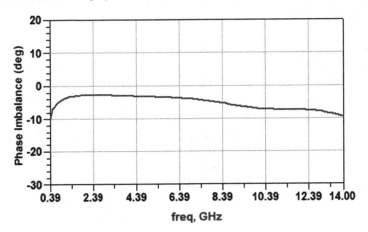

Fig. 4.4 Simulated phase imbalance for the proposed linearized CMOS distributed power splitter

is less than 1 dB over the band as shown in Fig. 4.5. The simulated phase S21 and phase S31 for the proposed linearized CMOS distributed power splitter is shown in Fig. 4.6.

4.2.2 CMOS Distributed Active Power Splitter Using Multiple-Gated Transistor Linearization

Active MOSFET devices can be modeled by nonlinear current and charge sources that depend on the device voltages [12, 13]. These nonlinear sources will give rise to distortion when driven with a modulated signal. The real MOSFET device output impedance is nonlinear and the mobility μ is not a constant but a function of the

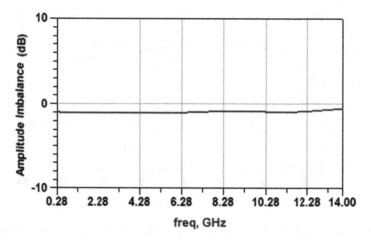

Fig. 4.5 Simulated amplitude imbalance for the proposed linearized CMOS distributed power splitter amplitude imbalance

Fig. 4.6 Simulated phase S21 and phase S31 for the proposed linearized CMOS distributed power splitter

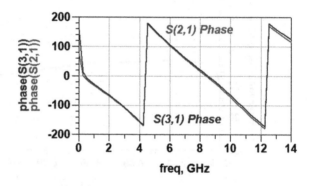

vertical and horizontal electric field. We may bias the active MOSFET device where the device behavior is more exponential. And there is also an internal feedback so when the input signal driven into the amplifier is increased, the output is also increased until a point where distortion products can no longer be ignored [12, 14]. No transistor is perfectly linear since the inherent nonlinearity of the diode junctions that comprise many of the active devices found in most amplifiers. The harmonics of the output signal are generated by nonlinearities of the MOSFET devices. The major three nonlinear elements of the MOSFET devices are nonlinear transconductance g_m, the device drain capacitance C_d and gate capacitance C_{gs} [12, 15]. The real MOSFET devices generate higher order distortion [12, 15, 16].

There are many linearization techniques to increase the linearity of the amplifier such as feedforward or multi-tanh [80–83] However they are better suited for differential circuits and hence consume more power and silicon area than linearization techniques applied to single-ended circuits. A simple linearization technique using

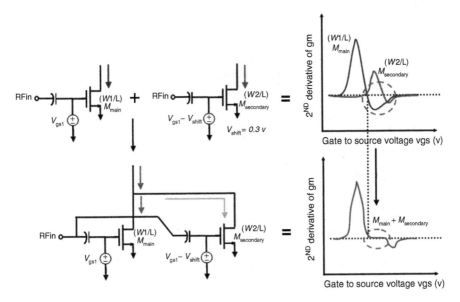

Fig. 4.7 Schematic illustration of multi-gated transistor linearization technique. The gate bias and the transistor size of the secondary transistor is chosen such that the negative second derivative of g_m peak of main transistor is canceled by the positive one of secondary transistor

multiple-gated common source transistors is used to linearize the single-ended CMOS distributed power splitter, where gate drive width and gate drive ($V_{gs} - V_{th}$) of each transistor are chosen to compensate for the nonlinear characteristic of the main transistor. Using Taylor series expansion, the drain current of a common source FET can be expressed as [84–86]

$$i_{ds} = I_{dc} + g_m v_{gs} + \frac{g_m'}{2} v_{gs^2} + \frac{g_m''}{6} v_{gs^3} + \dots \tag{4.4}$$

where g_m' and g_m'' are the first and second transconductance derivatives, respectively, with respect to the gate to source voltage. The negative g_m'' of the main transistor can be canceled by the positive g_m'' of the secondary transistor which is biased at a smaller gate drive as can be depicted in Fig. 4.7. The amount g_m'' compensation can be chosen by adjusting the width of secondary transistor. The compensated flat region of the overall transfer characteristic curve of both main and secondary transistor can be extended farther with proper bias voltage and transistor size to synthesize an overall transfer function with reduced nonlinearities [78, 79].

In this section, the multiple-gated transistor linearization technique is applied to the single-ended linearized CMOS distributed active power splitter design. The tuning of the power series nonlinear coefficient, transconductance g_m'', and its effect on IIP3 is shown in Figs. 4.8 and 4.9. As can be seen in Figs. 4.8 and 4.9 with proper

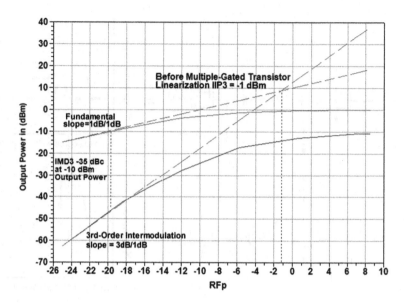

Fig. 4.8 ADS simulation of IIP3 before CMOS distributed active power splitter multiple-gated transistor linearization

bias voltage and transistor sizes the third-order nonlinearity can be canceled. Both IIP3 is improved by 8.5 dB and the third order intermodulation (IMD3) is improved by 10 dBc improvement, at output power of −10 dBm, can be seen in Fig. 4.9.

A common way to characterize the non-linear amplitude distortion of an amplifier under a two tone input is using the third order intercept point. This is the operating point where the power in the fundamental and the power in the third order intermodulation product are the same. In the linear region, third-order products increase by 3 dB for every 1 dB increase of input power, while output power increases by 1 dB. The output power where the two would intersect is the third order intercept point, and is a good indicator of the linearity of an amplifier

The design of a fully-integrated linearized CMOS distributed active power splitter that allows for broadband distortion reduction is presented in this section. Simulation results has yielded a peak S_{21} and S_{31} peak power gain of 7.8 dB and then rolls off to a unity gain bandwidth of 10.5 GHz with less than 5° phase imbalance and amplitude imbalance of less than 0.9 dB over the band. The simulation results show an 8.5 dB IIP3 improvement and a 10 dBc improvement at output power of −10 dBm and it has broadband signal transmission gain that compensates loss. The proposed fully-integrated design eliminates the need for off-chip discrete components and is suitable for ultra-wideband wireless transceiver applications.

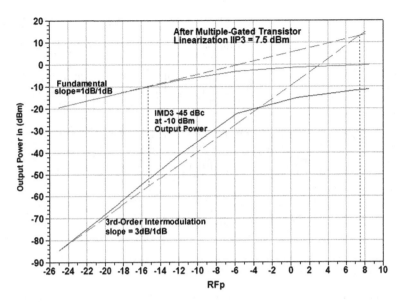

Fig. 4.9 ADS simulation of IIP3 after CMOS distributed active power splitter multiple-gated transistor linearization with an 8.5 dB IIP3 improvement and a 10 dBc IMD3 improvement at output power of −10 dBm

4.3 Linearized CMOS Distributed Matrix Amplifier Architecture

The interest in high-level integration and multi-functional subsystems motivates the development of broadband amplifiers. Distributed amplifiers have become the leading candidate for ultra-wideband amplification at microwave frequencies. A major design challenge for ultra-wideband amplifiers is the stringent linearity requirement over a wide bandwidth range, due to large numbers of in-band interferences in ultra-wideband wireless communication systems. Distributed Matrix amplifiers find many applications in broadband microwave applications and wireless communications [87–90]. A fully-integrated CMOS linearized interleaved distributed 2 × 3 matrix amplifier employing active post distortion and optimum gate bias linearization technique is presented in this section. The proposed CMOS linearized interleaved distributed 2 × 3 matrix amplifier incorporates a third-order intermodulation distortion IM3 suppression technique with simple circuit implementation and negligible extra power consumption that is well suited for integrated wireless broadband transceivers.

To apply the additive and multiplicative gain mechanism simultaneously for optimum amplifier performance, a matrix architecture for distributed amplifiers has been employed for circuit implementations [59, 91, 92]. Distributed matrix amplifier is a type of amplifier that combines the additive and the multiplicative

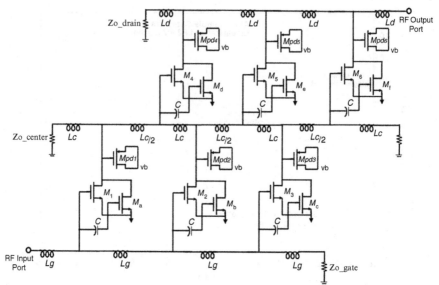

Schematic of Linearized CMOS interleaved distributed 2 x 3 matrix amplifier incorpating post distortion and optimum gate bias linearization technique.

Fig. 4.10 Schematic topology of the proposed fully-integrated linearized CMOS interleaved distributed 2 × 3 matrix amplifier employing active post distortion and optimum gate bias linearization technique

principles in one and the same module. It is possible to achieve both additive and multiplicative amplification by having two distributed amplifiers in a way that resembles the distribution of elements in an array therefore the name matrix amplifier as shown in Fig. 4.10. This two-tier matrix amplifier, which consists of a rectangular array of CMOS active devices, is composed of three periodically loaded artificial transmission lines: input line, central line, and output line. The input signal is amplified from one tier to the next before the final output signal leaves the uppermost drain-line output. This means that both multiplicative gain (one transistor output feeding the next's input) and additive gain (the summing of transistor outputs on the same drain-line) are provided within the single module. The structure is quite compact and allows circuits with higher gain per unit area as compared with conventional distributed amplifiers [21,64,66,73,93]. Figure 4.10 shows a schematic of a interleaved distributed 2 × n matrix amplifier and its equivalent circuit.

4.3.1   CMOS Distributed 2 × 3 Matrix Amplifier with Interleaved Distributed Loading Technique

In conventional distributed matrix amplifier, heavy loading on the central transmission line limits the bandwidth performance. Introducing an interleaved distributed loading technique would reduce the loading on the central artificial transmission line thus extending the bandwidth. The output loading of the first tier and the input loading of the second tier are separated by an equivalent inductance Lc in the proposed circuit topology as shown in Fig. 4.10. With the distributed loading technique, a significant performance improvement in terms of the bandwidth of the matrix amplifier can be achieved.

The central transmission line frequency f_c and characteristic impedance Z_{oc} for a conventional non-interleaved distributed matrix amplifier can be defined as [75]

$$f_{c,center} = \frac{1}{\pi \sqrt{L_c \left(C_{d2} + C_{g3} \right)}} \tag{4.5}$$

$$Z_{oc} = \frac{C_{g1}}{C_{d2} + C_{g3}} Z_o \tag{4.6}$$

where C_d is the drain capacitance and C_g is the gate capacitance. The transmission line central frequency f_c and characteristic impedance Z_{oc} for the interleaved distributed matrix amplifier with distributed loading technique can be defined as [17]

$$f_{c,center} = \frac{1}{\pi \sqrt{L_c C_{d2}}} = \frac{1}{\pi \sqrt{L_c C_{g3}}} \tag{4.7}$$

$$Z_{oc} = \frac{C_{g1}}{C_{d2}} Z_o = \frac{C_{g1}}{C_{g3}} Z_o \tag{4.8}$$

where C_d is the artificial transmission line drain capacitance and C_g is the gate capacitance. The conventional non-interleaved central artificial transmission lines of the matrix amplifier suffer from both gate and drain losses. The interleaved distributed matrix amplifier with distributed loading technique has a central transmission line frequency f_c equation with less capacitance and thus exhibit higher operational bandwidth, however it trades of with the small-signal gain of the amplifier.

The linearized CMOS interleaved distributed 2×3 matrix amplifier with distributed loading technique exhibit a 7.1 dB small signal S_{21} power gain peak as shown in Fig. 4.11 and rolls off to a unity gain bandwidth of 16 GHz with less than -10 dB return loss and isolation S_{12} of less than -45 dB over the band.

4.3.2   Proposed CMOS Interleaved Distributed 2 × 3 Matrix Amplifier with Post Distortion and Gate Optimum Bias Linearization Technique

We may bias the active MOSFET device where the device behavior is more exponential. The nonlinear MOSFET device sources will give rise to distortion when driven with a modulated signal. When the input signal driven into the amplifier is increased, the output is also increased until a point where distortion products can no longer be ignored [12, 14]. No transistor is perfectly linear since the inherent nonlinearity of the diode junctions that comprise many of the active devices found in

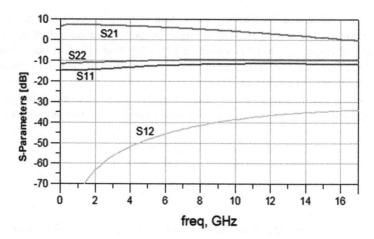

Fig. 4.11 Simulated output power S21 peaks at 7.1 dB and S-parameters for the proposed linearized CMOS interleaved distributed 2 × 3 matrix amplifier

most amplifiers. The harmonics of the output signal are generated by nonlinearities of the MOSFET devices. The major three nonlinear elements of the MOSFET devices are nonlinear transconductance g_m, the device drain capacitance C_d and gate capacitance C_{gs} [12,15]. The real MOSFET devices generate higher order distortion [12, 15, 16].

In this section, an interleaved distributed matrix amplifier with active post distortion and optimum gate bias distortion cancellation capability which is highly suitable for monolithic integration is presented [94]. The principle of active post distortion is to eliminate the considerable contribution to the overall amplifier distortion originating from the transconductance g_m nonlinearity. By introducing a linearizer after the input gain stage amplifier, the voltage of the MOS varactor is adjusted to synthesize a transfer function with reduced nonlinearities. The active post distortion linearizer compensates for the transconductance nonlinearities of the input gain stage amplifier producing a linear overall performance and the third-order nonlinearity is canceled as can be seen from Figs. 4.12 to 4.19.

The drain and gate transmission-line inductors have an inductance value of 1.3 nH and the drain transmission-line m-derived inductance is 100 pH with 50 Ohms terminations. The device dimensions $[M_1, M_6]$ (W/L) have a width equal to 48 μm with all devices having L minimum channel length of 120 nm. The device dimensions $[M_a, M_f]$ (W/L) have a width equal to 24 μm with all devices having L minimum channel length of 120 nm. The device dimensions M_{pd} (W/L) have a width equal to 40 μm with minimum channel length L of 120 nm.

ADS Simulation of IIP3 before CMOS distributed 2 × 3 matrix amplifier active post distortion and optimum gate bias linearization is shown in Fig. 4.14. Simulation of IIP3 after CMOS distributed 2 × 3 matrix amplifier active post distortion and optimum gate bias linearization with a 9 dB and a 18 dBc improvement at output power of −10 dBm is shown in Fig. 4.15.

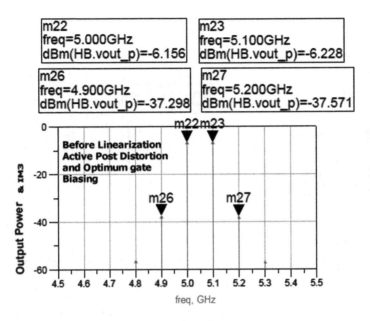

Fig. 4.12 Simulated Two-Tone before linearization at 5 GHz

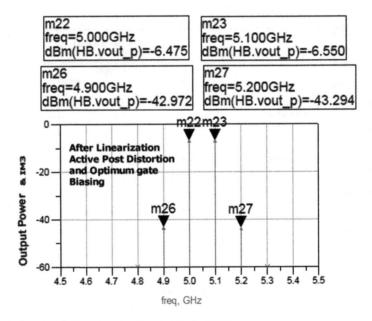

Fig. 4.13 Simulated Two-Tone after linearization at 5 GHz

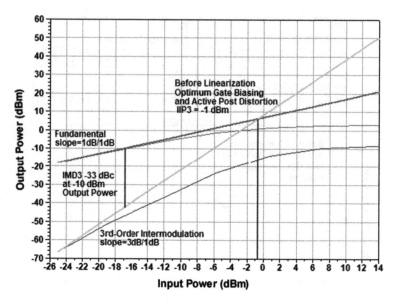

Fig. 4.14 ADS Simulation of IIP3 before CMOS distributed 2 × 3 matrix amplifier active post distortion and optimum gate bias linearization

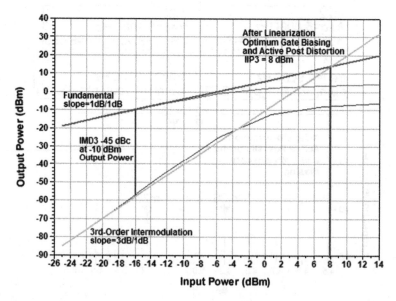

Fig. 4.15 ADS Simulation of IIP3 after CMOS distributed 2 × 3 matrix amplifier active post distortion and optimum gate bias linearization with a 9 dB and a 18 dBc improvement at output power of −10 dBm

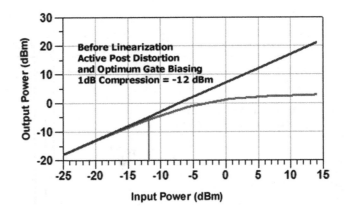

Fig. 4.16 Simulated 1-dB gain compression point before linearization for the proposed linearized CMOS distributed 2 × 3 matrix amplifier

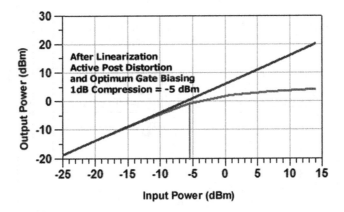

Fig. 4.17 Simulated 1-dB gain compression point after linearization for the proposed linearized CMOS interleaved distributed 2 × 3 matrix amplifier

A common way to characterize the non-linear amplitude distortion of an amplifier is with the 1 dB compression point. This is the point at which the amplifier gain drops 1 dB below the linear gain. Simulated 1-dB gain compression point before linearization for the proposed linearized CMOS distributed 2 × 3 matrix amplifier is shown in Fig. 4.16. Simulated 1-dB gain compression point after linearization for the proposed linearized CMOS interleaved distributed 2 × 3 matrix amplifier in Fig. 4.17.

Simulated Two-Tone before active post distortion linearization only at 5 GHz for the proposed linearized CMOS interleaved distributed 2 × 3 matrix amplifier is shown in Fig. 4.18. Simulated Two-Tone after active post distortion linearization only at 5 GHz for the proposed linearized CMOS interleaved distributed 2×3 matrix amplifier in Fig. 4.19. The concept of optimum gate bias linearization technique is that the negative third-order transconductance g_m peak of one transistor can

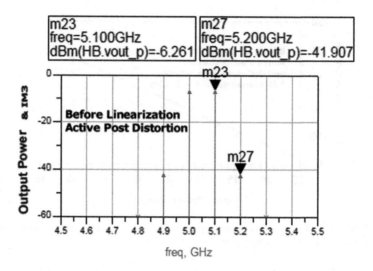

Fig. 4.18 Simulated Two-Tone before active post distortion linearization only at 5 GHz for the proposed linearized CMOS interleaved distributed 2 × 3 matrix amplifier

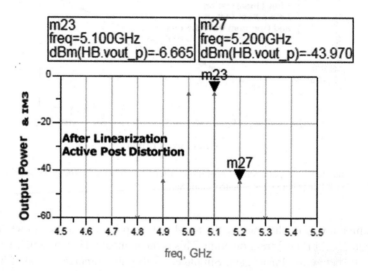

Fig. 4.19 Simulated Two-Tone after active post distortion linearization only at 5 GHz for the proposed linearized CMOS interleaved distributed 2 × 3 matrix amplifier

be canceled by the positive peak of an auxiliary MOSFET with an optimum gate voltage offset bias so that the magnitude of the its positive third-order transconductance g_m peak is equal to the negative peak of the master transistor [25]. The combined transistors yield a third-order transconductance nonlinearity near-zero and in turn improves its linearity. It can be concluded from simulations that tuning the linearized CMOS distributed 2 × 3 matrix amplifier using active post distortion and optimum gate bias improves its broadband linearity and in turn

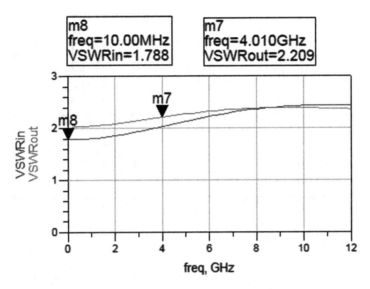

Fig. 4.20 The VSWR for the proposed linearized CMOS interleaved distributed 2 × 3 matrix amplifier

improves its dynamic range as well. As can be seen in Figs. 4.16 and 4.17 by shifting the bias point and superposing the device characteristics the third-order nonlinearity can be canceled and 1-dB gain compression point is improved as well.

The varactor-based active post distortion enables one to tune the IM3 suppression as can be seen in Figs. 4.18 and 4.19, it provides repeatable electronically tunable distortion cancellation at various combination of varactor voltages. The VSWR voltage standing wave ratio for the proposed Linearized CMOS Interleaved Distributed 2 × 3 Matrix Amplifier is shown in Fig. 4.20.

4.4 Linearized CMOS Distributed Paraphase Amplifier

Broadband paraphase amplifiers find many applications in microwave applications and wireless communications. Circuits which are suitable for fully monolithic integration in cost effective CMOS technology are of particular interest [95]. Sokolov and Williams [55] have reported on a narrowband MMIC GaAs paraphase amplifier at 8.5 GHz and Levent-Villegas [96] has reported on another narrowband GaAs paraphase amplifier over the range of 9–11 GHz.

A fully-integrated CMOS linearized distributed paraphase amplifier (unbalanced input, balanced output) is presented in this section. The CMOS linearized distributed paraphase amplifier or linearized active balun will provide two outputs, 180° out of phase over a wide microwave band with minimal phase and amplitude imbalance. Compared to passive balun circuits, the linearized CMOS distributed paraphase amplifier's advantage is that it provides gain and allows for broadband distortion

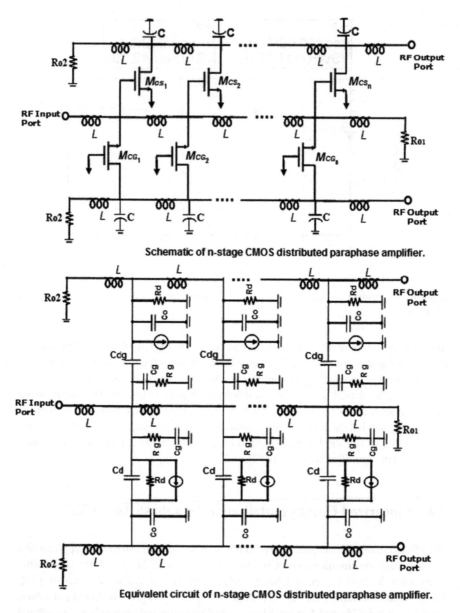

Schematic of n-stage CMOS distributed paraphase amplifier.

Equivalent circuit of n-stage CMOS distributed paraphase amplifier.

Fig. 4.21 Block diagram illustration of CMOS distributed paraphase amplifier

cancellation. The proposed CMOS linearized distributed paraphase amplifier makes up for the inherent 3 dB loss of balun and be used to drive balanced antennas and wideband phase shifters.

The wideband distributed paraphase amplifier is based on the concept of the distributed amplifier [27, 84, 97]. Figure 4.21 shows a schematic of an n-stage distributed paraphase amplifier and its equivalent circuit. It consists of n-stage distributed FET amplifiers. One is a distributed common-source amplifier, the other is a distributed common-gate amplifier. These two amplifiers share a common input transmission line. A simple circuit model for the distributed paraphase amplifier is shown Fig. 4.21.

The power gain of the common-gate section of the distributed paraphase amplifier shown in Fig. 4.21 is given by [97]

$$
G_{CG} = \frac{G_m^2 R_{og} R_{od} \left(1 + \frac{\sqrt{\left[1 + \left(\varpi/\varpi_g\right)^2\right]}}{G_m R_d}\right)^2}{4\left[1 + \left(\varpi/\varpi_g\right)^2\right]^{1/2}\left[1 - \left(\varpi/\varpi_c\right)^2\right]\left(1 + \frac{Z_{d\pi}}{2R_d}\right)^2}
$$

$$
\times \frac{\sinh^2\left[\frac{n}{2}\left(\alpha_d - \alpha_g\right)\right]e^{-n\left(\alpha_d + \alpha_g\right)}}{\sinh^2\left[\frac{1}{2}\left(\alpha_d - \alpha_g\right)\right]} \tag{4.9}
$$

where G_m is the transconductance and α_d and α_g are the attenuation per section and R_{od} and R_{og} are the characteristic resistance and ϖ_d and ϖ_g are of the drain and gate line respectively.

The power gain of the distributed common-source amplifier in the distributed paraphase amplifier shown in Fig. 4.21 is given by [29,97]

$$
G_{CS} = \frac{G_m^2 R_{og} R_{od}}{4\left[1 + \left(\varpi/\varpi_g\right)^2\right]^{1/2}\left[1 - \left(\varpi/\varpi_c\right)^2\right]}
$$

$$
\times \frac{\sinh^2\left[\frac{n}{2}\left(\alpha_d - \alpha_g\right)\right]e^{-n\left(\alpha_d + \alpha_g\right)}}{\sinh^2\left[\frac{1}{2}\left(\alpha_d - \alpha_g\right)\right]} \tag{4.10}
$$

where R_{od} and R_{og} are the characteristic resistance and G_m is the transconductance.

The amplitude imbalance of the distributed paraphase amplifier can be expressed as [97]

$$
\Delta A\,(dB) = 20\log \frac{1 + \frac{1}{G_m R_d}}{1 + \frac{Z_{d\pi}}{2R_d}} + 20\log \frac{\cosh\left(\frac{n}{2}\frac{\alpha_d}{\alpha_o}\right)}{\cosh\left(\frac{1}{2}\frac{\alpha_d}{\alpha_o}\right)}
$$

$$
- 10\log\left[\exp\left(n\frac{\alpha_d}{\alpha_o}\right)\right] \tag{4.11}
$$

where G_m is the transconductance. The first term results from the gain difference between the two stages while the latter two terms stem from the difference in attenuation on the two output lines.

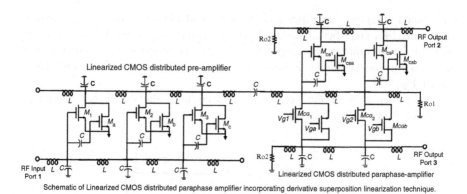

Schematic of Linearized CMOS distributed paraphase amplifier incorporating derivative superposition linearization technique.

Fig. 4.22 Schematic topology of the proposed fully-integrated linearized CMOS distributed paraphase amplifier

The phase imbalance of more practical concern is given by [97]

$$\delta\phi = \arctan \frac{\varpi C_d}{\frac{1}{R_d} + G_m} - \arctan \frac{\varpi C_d}{\frac{1}{R_d} + \frac{2}{Z_{d\pi}}} \qquad (4.12)$$

where the G_m is the device transconductance. The design parameters such as the output line impedance is dictated by circuit constraints the device can be chosen to minimize the phase imbalance.

4.4.1 Amplitude and Phase Imbalance of Linearized CMOS Distributed Paraphase Amplifier

The inherent broadband characteristics of the distributed amplifier is applied to a linearized CMOS distributed paraphase design. In the amplifier a distributed common-gate amplifier and a distributed common-source amplifier are fed by a single input transmission line. The characteristic low pass response of the distributed amplifier then yields two RF outputs from one RF input signal, the amplitudes of both RF output signals are equal and their phase difference is 180° over a wide band [78, 79, 98]. Figure 4.22 shows the schematic of the proposed linearized CMOS distributed paraphase amplifier. A distributed pre-amplifier stage is added which not only supplies gain but serves as an active impedance transformer.

The lumped transmission lines tend to provide a constant input and output resistance over a wide passband. As a result, the phase difference between the output of the common-gate amplifier and the output of the common-source amplifier is theoretically 180°, independent of the frequency, if the phase velocities of the signals on the three transmission lines are identical. The voltage gains of the

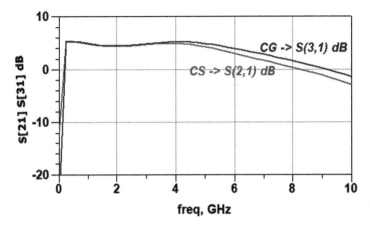

Fig. 4.23 Simulated output power S21 common-source stage port and S31 common-gate stage port peaks at 5.3 dB for the proposed linearized CMOS distributed paraphase amplifier

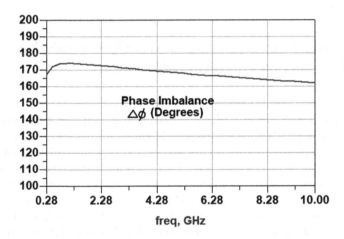

Fig. 4.24 Simulated phase imbalance for the proposed linearized CMOS distributed paraphase amplifier with less than 10° phase imbalance

common-source amplifier and the common-gate amplifier are almost the same with nearly equal output power as can be seen in Fig. 4.23 and as a result, a wideband CMOS distributed paraphase amplification is obtained.

The linearized CMOS distributed paraphase amplifier exhibiting an 5.3 dB small signal S_{21} power gain peak as shown in Fig. 4.23 and rolls off to a unity gain bandwidth of 8.2 GHz with less than 10° phase imbalance as shown in Fig. 4.24 and amplitude imbalance of less than 1.26 dB over the band as shown in Fig. 4.25.

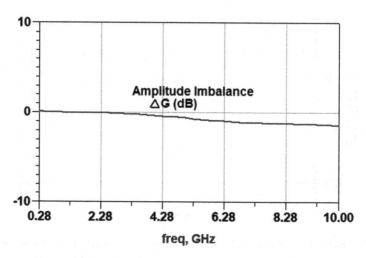

Fig. 4.25 Simulated amplitude imbalance for the proposed linearized CMOS distributed para-phase amplifier amplitude imbalance of less than 1.26 dB

4.4.2   CMOS Distributed Paraphase Amplifier Employing Derivative Superposition Linearization

The principle of derivative superposition is to eliminate the considerable contribution to the overall amplifier distortion originating from the transconductance G_m nonlinearity [99, 100]. By running different sized transistors in parallel the direct current operating point of each single device is adjusted to synthesize a transfer function which does not generate any nonlinearities. The drain and gate transmission-line inductors have an inductance value of 1.2 nH and the drain transmission-line m-derived inductance is 110 pH with 50 Ohms terminations. The device dimensions $[M_1, M_3]$ (W/L) and $[M_{cs1}, M_{cs2}]$ (W/L) have a width equal to 48 μm with all devices having L minimum channel length of 120 nm. The device dimensions $[M_a, M_c]$ (W/L) and $[M_{csa}, M_{csb}]$ (W/L) have a width equal to 24 μm with all devices having L minimum channel length of 120 nm.

The derivative superposition technique [65] is based on the drain current dependence versus V_{gs}. When considering the related third order Taylor coefficient of this current gm3, which represents the third order nonlinearity, this bias dependence results in magnitude and phase reversals for gm3 in the threshold region. The generalized circuit topology to be used with derivative superposition is shown in Fig. 4.26.

In this section the derivative superposition linearization technique is applied to the linearized CMOS distributed paraphase design. The tuning of the power series nonlinear coefficient, transconductance gm, and its effect on IIP3 is shown in Figs. 4.27 and 4.28. It can be concluded from simulations that tuning the linearized CMOS distributed paraphase amplifier using derivative superposition improves

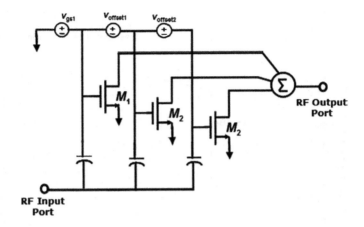

Schematic of Linearized CMOS common source amplifier incorporating derivative superposition linearization technique.

Fig. 4.26 The derivative superposition linearization approach. The overall linearized transconductance is composed out of several gate bias shifted common source FET devices, which are current summed at their outputs

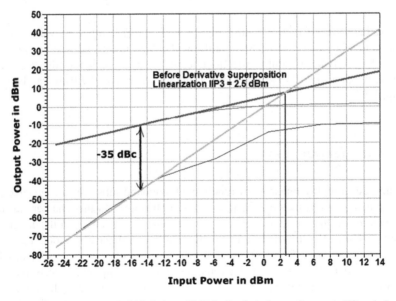

Fig. 4.27 ADS simulation of IIP3 before CMOS distributed paraphase amplifier derivative superposition linearization

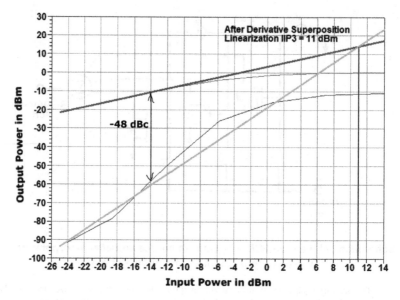

Fig. 4.28 ADS simulation of IIP3 after CMOS distributed paraphase amplifier derivative superposition linearization with a 8.5 dB and a 13 dBc improvement at output power of −10 dBm

its broadband linearity and in turn improves its dynamic range as well. As can be seen in Figs. 4.27 and 4.28 by shifting the bias point and superposing the device characteristics the third-order nonlinearity can be canceled. Both 8.5 dB IIP3 improvement and 13 dBc improvement, at output power of −10 dBm, can be seen in Fig. 4.28.

4.5 Chapter Summary

The interest in highly-linear multi-functional broadband subsystems motivated the development of linearized distributed circuit applications. In this chapter, various applications of linearized distributed circuit functions such as fully-integrated CMOS linearized distributed active power splitter (unbalanced input, balanced output) and a linearized CMOS distributed matrix amplifier were presented. Also the linearized CMOS distributed paraphase amplifier employing derivative superposition linearization was presented as well. The proposed fully-integrated linearized distributed circuit designs are suitable for highly-linear broadband wireless transceiver applications.

Chapter 5
Linearized CMOS Distributed Bidirectional Amplifier with Cross-Coupled Compensator

5.1 Introduction

In this chapter, we demonstrate a fully-integrated fully-differential linearized CMOS distributed bidirectional amplifier that achieves large IMD3 distortion reduction over broadband frequency range for both RF paths. The drain and gate transmission-lines were stagger-compensated. Reducing the DA IM3 distortion by mismatching the gate and drain LC delay-line ladders. A CMOS cross-coupled compensator transconductor is proposed to enhance the linearity of the DA gain cell with a varactor-based active post nonlinear drain capacitance compensator for wider linearization bandwidth.

5.2 Linearized CMOS Distributed Bidirectional Amplifier Circuit Design Analysis

For wireless base station transmitters in UWB communications such as bidirectional UWB RoF transmission, high linearity and broadband bandwidth are some of the primary system requirements. Linear broadband amplification is required in order to reduce signal spectral regrowth. The wide bandwidth of the DA structure makes it an attractive component for use in systems requiring broadband operation. A detailed paper by Ginzton et al. [56] implemented Percival's patented concept of distributed amplification [74]. Conventional DAs use low-pass π-sections to form an artificial transmission-line topology. Amplification gain stages are connected so that output currents are combined in an additive manner at the output terminal. The advantages of a DA topology are its wide bandwidth, flat gain and compact size circuit size. A three-stage CMOS DA with MOSFET devices is shown in Fig. 5.1. The major three nonlinear elements of the MOSFET devices are g_m, C_d and C_{gs}. An ideal broadband amplifier should be a totally linear device. But in real world, amplifiers are only linear within certain practical limits. When the input signal driven into

Z. El-Khatib et al., *Distributed CMOS Bidirectional Amplifiers: Broadbanding and Linearization Techniques*, Analog Circuits and Signal Processing, DOI 10.1007/978-1-4614-0272-5_5, © Springer Science+Business Media New York 2012

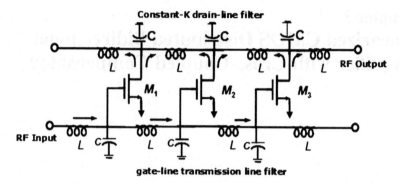

Fig. 5.1 A three-stage CMOS DA with MOSFET devices

the amplifier is increased, the output is also increased until a point where distortion products can no longer be ignored. The harmonics of the output signal are generated by nonlinearities of the MOSFET devices. The nonlinear element transconductance g_m is a function of V_{gs} and can be expressed by power series with coefficients g_{m1}, g_{m2} and g_{m3} as in (5.1).

The characteristic of the nonlinear transconductance can be approximated by a three-term power series expansion for each MOSFET device in Fig. 5.1 [13, 16, 36, 83, 101]

$$i_d = g_{m1} v_{gs} + g_{m2} v_{gs}^2 + g_{m3} v_{gs}^3 \tag{5.1}$$

As the input signal increases the instantaneous value of g_m changes with the level of the input signal and the amplifier operating into its nonlinear region. We can write v_{gs} as [13, 84]

$$v_{gs} = V \cos(\omega t) \tag{5.2}$$

where V is the amplitude signal and substituting (5.2) into (5.1), the expression for the current at frequency ω is given by [15, 85, 102–104]

$$i = g_{m1} V \cos(\omega t) + \frac{3}{4} g_{m3} V^3 \cos(\omega t) \tag{5.3}$$

In the three-stage CMOS DA shown in Fig. 5.1, the fundamental component of the output current from each stage can be defined as [15, 102–104]

$$i_1 = g_{m1} V \cos[\omega] e^{-j1\beta_g} \tag{5.4}$$

$$i_2 = g_{m1} V \cos[\omega] e^{-j2\beta_g} \tag{5.5}$$

$$i_3 = g_{m1} V \cos[\omega] e^{-j3\beta_g} \tag{5.6}$$

$$i_k = g_{m1} V \cos[\omega] e^{-jk\beta_g} \tag{5.7}$$

where β_g is the phase constant of the gate line.

The total fundamental output current delivered to the load for all three-stages can be found as [15, 102–104]

$$I_{linear} = \frac{1}{2}\left(i_1 e^{-jn\beta_d} + i_2 e^{-j(n-1)\beta_d} + i_3 e^{-j(n-2)\beta_d}\right) \tag{5.8}$$

where β_d is the phase constant of the drain line and n is a constant representing the number of stages.

In the three-stage CMOS DA as shown in Fig. 5.1, the third-order intermodulation IM3 component of the output current from each stage can be defined as

$$i_{m1} = \frac{3}{4} g_{m3} V^3 cos[\omega] e^{-j3\beta_g} \tag{5.9}$$

$$i_{m2} = \frac{3}{4} g_{m3} V^3 cos[\omega] e^{-j6\beta_g} \tag{5.10}$$

$$i_{m3} = \frac{3}{4} g_{m3} V^3 cos[\omega] e^{-j9\beta_g} \tag{5.11}$$

$$i_{mk} = \frac{3}{4} g_{m3} V^3 cos(\omega t) e^{-j3k\beta_g} \tag{5.12}$$

The corresponding total IM_3 output current delivered to the load for all three-stages can be found as

$$I_{IM3} = \frac{1}{2}\left(i_{m1} e^{-jn\beta_d} + i_{m2} e^{-j(n-1)\beta_d} + i_{m3} e^{-j(n-2)\beta_d}\right) \tag{5.13}$$

Dividing (5.8) by (5.13) yields

$$I_{linear}/I_{IM3} = \frac{4}{3} e^{6Re[j\beta_g]} \left| \frac{\left(e^{2j\beta_d} + e^{2j\beta_g} + e^{j(\beta_d+\beta_g)}\right) g_{m1}}{\left(e^{2j\beta_d} + e^{6j\beta_g} + e^{j(\beta_d+3\beta_g)}\right) g_{m3} V^2} \right| \tag{5.14}$$

It can be seen from (5.14) as well that the carrier power to third-order intermodulation C/IM_3 is dependent on the power series nonlinear coefficients g_{m1} and g_{m3} and that C/IM_3 can be improved by reducing the g_{m3} nonlinearity.

Equation 5.14 can be extended to an n-stage CMOS DA. The total fundamental output current delivered to the load for n-stages can be found as

$$In_{linear} = \frac{1}{2} \sum_{r=1}^{n} g_{m1} V cos[\omega t] e^{-jr\beta_g} e^{-j(n-r+1)\beta_d} \tag{5.15}$$

The total fundamental output current delivered to the load for n-stages can be simplified as

$$In_{linear} = \frac{e^{-jn\beta_d}\left(-1 + \left(e^{j(\beta_d-\beta_g)}\right)^n\right) g_{m1} V Cos[t\omega]}{2\left(e^{j\beta_d} - e^{j\beta_g}\right)} \tag{5.16}$$

The corresponding total IM_3 output current for all n-stages flowing through the drain load can be defined as

$$In_{IM3} = \frac{1}{2} \sum_{r=1}^{n} \frac{3}{4} g_{m3} V^3 Cos[\omega t] e^{-j3r\beta_g} e^{-j(n-r+1)\beta_d} \tag{5.17}$$

The total IM_3 output current for n-stages flowing through the drain load can be simplified as

$$In_{IM3} = \frac{3e^{-jn\beta_d} \left(-1 + \left(e^{j\left(\beta_d - 3\beta_g\right)}\right)^n\right) g_{m3} V^3 Cos[t\omega]}{8\left(e^{j\beta_d} - e^{3j\beta_g}\right)} \tag{5.18}$$

Dividing (5.16) by (5.18) yields

$$In_{linear}/In_{IM3} = \frac{4}{3} \frac{\left(e^{j\beta_d} - e^{3j\beta_g}\right)\left(-1 + \left(e^{j\left(\beta_d - \beta_g\right)}\right)^n\right) g_{m1}}{\left(e^{j\beta_d} - e^{j\beta_g}\right)\left(-1 + \left(e^{j\left(\beta_d - 3\beta_g\right)}\right)^n\right) g_{m3} V^2} \tag{5.19}$$

From (5.16) the overall fundamental filter frequency response is expressed as

$$F(s)_{fund} = \frac{e^{-jn\beta_d}\left(-1 + \left(e^{j\left(\beta_d - \beta_g\right)}\right)^n\right)}{2\left(e^{j\beta_d} - e^{j\beta_g}\right)} \tag{5.20}$$

From (5.18) overall IM3 filter frequency response is expressed as

$$F(s)_{im3} = \frac{3e^{-jn\beta_d}\left(-1 + \left(e^{j\left(\beta_d - 3\beta_g\right)}\right)^n\right)}{8\left(e^{j\beta_d} - e^{3j\beta_g}\right)} \tag{5.21}$$

where the phase constant of the gate line is given by $\beta_g = \omega\sqrt{L_g.C_g}$ and $\beta_d = \omega\sqrt{L_d.(C_d + C_{var})}$ is the phase constant of the drain line.

The characteristic impedance of the drain line is given by $Z_{o\_d} = \sqrt{\frac{C_d}{L_d}}$ and the drain cut-off frequency is given by $\omega_c = \frac{1}{\sqrt{L_d C_d}}$.

Designing a staggered bidirectional DA with mismatched time-delay for the drain and gate line can be achieved by mismatching the drain and gate line capacitance. A simulation of a staggered drain and gate transmission-lines with mismatched time-delay to filter out the IM3 distortion improving the C/IM_3 in broadband CMOS bidirectional DAs is shown in Fig. 5.2a. Further increase in the time-delay mismatch will shift the IM_3 attenuation frequency response inside the in-band of the DA cutoff frequency as can be seen in Figs. 5.2b and 5.3. Thus, having a staggered distributed structure helps in improving the carrier power to third-order intermodulation distortion power ratio C/IM_3 in broadband CMOS amplifiers [30].

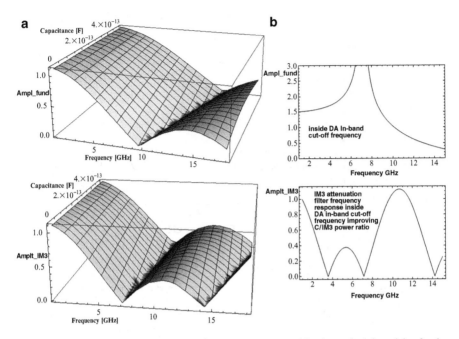

Fig. 5.2 (**a**) Designing a staggered bidirectional DA structure with mismatched time-delay for the drain and gate line will shift the IM3 attenuation frequency response inside the in-band of the DA cut-off frequency and in turn out the IM3 distortion. (**b**) Further increase in the time-delay mismatch will shift the IM_3 attenuation frequency response inside the in-band of the DA cutoff frequency

5.3   CMOS Cross-Coupled Compensator Transconductor as DA Gain Cell for Linearity Improvement and Enhanced Tunability

In radio transceivers, a differential pair transconductor is often used to convert the RF input voltage signal to a current that is either amplified or frequency translated [32, 34]. External linearization circuitry can be added to the differential pair amplifier in order to compensate for the transconductance nonlinearity [58]. This allows the nonlinear amplifier to be used to amplify signals utilizing spectrally efficient linear modulation techniques without causing interference [33, 34]. Many techniques have been developed to improve the linearity of the basic differential pair transconductor such as source degeneration, Caprio's cross-quad [105] and Quinn's cascomp transconductor [106]. However, their second derivative device transconductor transistor parameter gm″ nulling is not sufficiently wide and their device transconductor parameter gm flatness is limited. Other published linearized BJT V-I converters such as proposed in [80] does not linearize the total transconductor drain current instead only linearizing the inner translinear loop current with

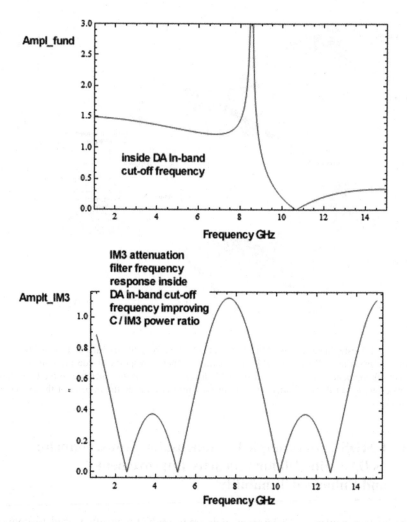

Fig. 5.3 More increase in time delay mismatch by increasing the drain and gate line capacitance mismatching will shift the IM3 attenuation frequency response even more inside the in-band of the distributed amplifier cut-off frequency further filtering out IM3 distortion

no on-chip tuning. The proposed linearized cross-coupled compensator transconductor achieves a wide tunable gm'' nulling and increase in device transconductor parameter gm flatness.

In this chapter, we propose a 0.13 μm RF CMOS highly-linear differential cross-coupled compensator transconductor that combines both cross-quad and cascomp linearization techniques with enhanced tunability [80]. The proposed cross-coupled compensator linearize the total drain current while maintaining a high input voltage swing range. It achieves a wide tunable second derivative device transconductor parameter gm'' nulling over 500 mV differential input signal and a tunable 50° increase in the device transconductor parameter gm flatness. The proposed linearized transconductor offers significant improvement in linearity for the use in highly-linear broadband high-frequency amplifiers [30]. The proposed cross-coupled compensator transconductor [99] maintains a high input voltage swing range due to the aid of the second translinear loop that is not implemented in the other published BJT V-I converter in [80].

The proposed differential CMOS cross-coupled compensator transconductance is implemented as shown in Fig. 5.4a. A source degeneration resistor R_b is placed across the outer cascode differential pair in between the identical current sources $Ibias_b$. Similarly a source degeneration resistor R_t is placed in between current source $Ibias_t$ inner cascode differential pair. To achieve a distortion-less V-I linear conversion of the transconductor total current ΔI_{out}, both bias current sources $Ibias_b$ and $Ibias_t$ are tuned to ensure both currents ΔI_t and ΔI_b become linearly proportional to the input voltage ΔV_{in}. Hence the proposed CMOS cross-coupled compensator total current $\Delta I_{out} = \Delta I_t + \Delta I_b$ is linearly proportional to the input voltage ΔV_{in} contrary to the linearized BJT transconductor in [107]. In the linearized BJT transconductor the inner translinear loop only considers the inner loop current ΔI_{inner} which becomes linearly proportional to the input voltage $\Delta V_{in} = (2Ree).\Delta I_{inner}$ and does not consider the total current ΔI_{out} of the transconductor.

Lets consider the cross-coupled compensator sum of voltages around the first translinear loop comprising the signal generator and gate-source voltages of (M_1, M_9, M_{10}, M_2) and the degeneration resistor R_t as shown in Fig. 5.4a. Ignoring body effects yields the following expression:

$$\Delta I_t = \frac{\Delta V_{in} - (V_{GS1} + V_{GS9} - V_{GS10} - V_{GS2})}{R_t} \tag{5.22}$$

Next, consider the sum of the voltages around the second translinear loop comprising the gate-source voltages of (M_3, M_7, M_8, M_4) and the degeneration resistor R_b as shown in Fig. 5.4a. Ignoring body effects yields the following expression:

$$\Delta I_b = \frac{(V_{GS3} + V_{GS7} - V_{GS8} - V_{GS4})}{R_b} \tag{5.23}$$

Adding both loops together yields the total current ΔI_{out_total}:

$$\Delta I_{out\_total} = \Delta I_t + \Delta I_b \tag{5.24}$$

$$\Delta I_{out\_total} = \frac{\Delta V_{in} - (V_{GS1} + V_{GS9} - V_{GS10} - V_{GS2})}{R_t}$$
$$+ \frac{(V_{GS3} + V_{GS7} - V_{GS8} - V_{GS4})}{R_b} \tag{5.25}$$

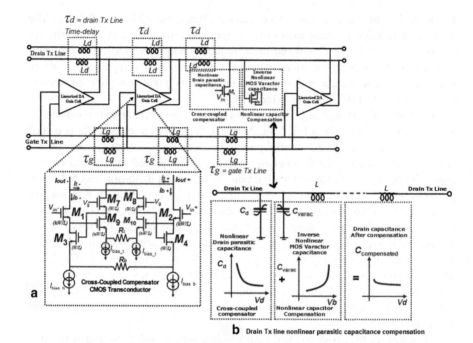

Fig. 5.4 (**a**) Proposed topology of the three-stage bidirectional distributed amplifier with CMOS cross-coupled compensator transconductor gain cells coupling the staggered drain and gate transmission-lines with nonlinear drain capacitance compensator. (**b**) A varactor-based active post nonlinear drain capacitance compensator in the CMOS distributed structure for wider linearization bandwidth

$$\Delta I_{out\_total} = \frac{\Delta V_{in}}{R_t} - \frac{(V_{GS1} + V_{GS9} - V_{GS10} - V_{GS2})}{R_t}$$
$$+ \frac{(V_{GS3} + V_{GS7} - V_{GS8} - V_{GS4})}{R_b} \tag{5.26}$$

For the total differential output current ΔI_{out} to achieve a linear relationship with the input differential voltage V_{in} and hence enhancing linearity, only when the following relationship is satisfied,

$$\frac{(V_{GS1} + V_{GS9} - V_{GS10} - V_{GS2})}{R_t} = \frac{(V_{GS3} + V_{GS7} - V_{GS8} - V_{GS4})}{R_b} \tag{5.27}$$

For (5.27) to be satisfied, the following (5.28) and (5.29) must be true:

$$(V_{GS3} - V_{GS1}) + (V_{GS10} - V_{GS8}) = 0 \tag{5.28}$$

$$(V_{GS2} - V_{GS4}) + (V_{GS7} - V_{GS9}) = 0 \tag{5.29}$$

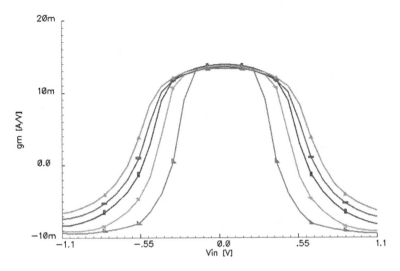

Fig. 5.5 CMOS cross-coupled compensator tunable gm flatness versus input signal

Equations 5.28 and 5.29 are met by adjusting both bias current sources $Ibias_t$ and $Ibias_b$. Having the device sizes with unequal dimensions, (5.28) and (5.29) can be satisfied by tuning both bias current source $Ibias_t$ and $Ibias_b$. Transistors with smaller device size (W/L) mean that their equivalent gate-source voltage V_{gs} changes faster. Transistors with larger device size (W/L) have their gate-source voltage V_{gs} change slower. When both bias current source $Ibias_t$ and $Ibias_b$ are adjusted, the cross-coupled compensator g_m is tuned achieving higher linearity. Hence the total differential output current ΔI_{out} achieves a linear relationship with the input differential voltage V_{in}.

To ensure circuit stability the device dimensions were optimized dimension of $[M_1,M_2]$ width (W/L) equal to 96 μm and $[M_3,M_4]$ (W/L) equal to 72 μm and $[M_7,M_8]$ (W/L) equal to 32 μm and $[M_9,M_{10}]$ (W/L) equal to 20 μm with all devices having L minimum channel length of 120 nm. Simulation results show that the proposed linearized transconductor achieves a tunable 50° increase in the device transconductor parameter gm flatness over wide differential input signal as shown in Fig. 5.5 and a wide second derivative device transconductor parameter gm″ nulling over 500 mV as shown in Fig. 5.6. The proposed transconductor [99] maintains a high input voltage swing range due to the aid of the second translinear loop ($M_3 - M_4$).

The distributed three-stage CMOS cross-coupled compensator transconductor structure is formed through adopting a low-pass artificial transmission line into a bidirectional DA as shown in Fig. 5.4. Each CMOS cross-coupled compensator transconductor stage is separated with series on-chip spiral inductors used to extend the operation frequency range. The currents from each stage are combined at the output terminal and therefore the third-order intermodulation IM3 distortion reduction is effective over a broad bandwidth.

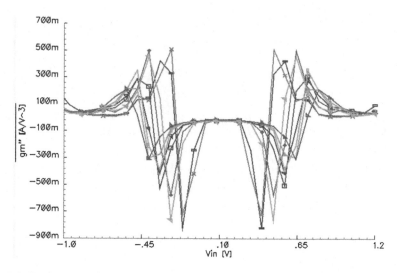

Fig. 5.6 Simulation result of CMOS cross-coupled compensator wide tunable gm'' nulling versus input signal

5.4 Effect of Nonlinear Drain Capacitance on DA Linearization Bandwidth

Another source of nonlinearity in CMOS bidirectional DAs is the transmission line nonlinear drain capacitance C_d. A nonlinear drain current is induced flowing out of the drain of an NMOS transistor. This nonlinearity can be reduced by introducing a parallel inverse nonlinearity at the transmission drain line at the output of the CMOS distributed structure to compensate for the drain capacitance of the active element as depicted in Fig. 5.4b. The PMOS varactor-based active PMOS nonlinear capacitance compensator tunes the nonlinear drain capacitance providing IM3 distortion cancellation at various varactor voltages. The transmission-line capacitance is part of the filter structure whose bandwidth is determined by the amount of transmission-line inductance and capacitance of the filter section. The transmission-line cut-off frequency f_c is widened when the transmission-line capacitance is reduced [84, 97]

$$f_c = \frac{1}{\pi\sqrt{LC}}. \tag{5.30}$$

The nonlinear element drain capacitance C_d is a function of v_{ds} and can be expressed by power series with coefficients C_{d0}, C_{d1} and C_{d2}

$$C_d = C_{d0} + C_{d1}V_{ds} + C_{d2}V_{ds}^2 \tag{5.31}$$

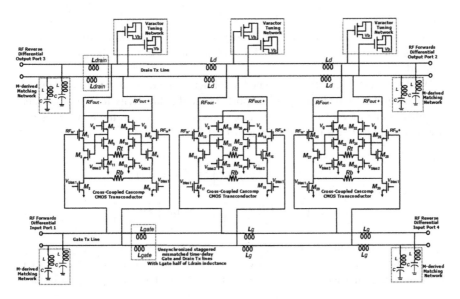

Fig. 5.7 The proposed schematic of the three-stage bidirectional distributed amplifier with CMOS cross-coupled compensator transconductor gain cells coupling the staggered drain and gate transmission-lines with nonlinear drain capacitance compensator

The nonlinear element C_{var} is a function of v_{ds} and can be expressed by power series with coefficients C_{var0}, C_{var1} and C_{var2}

$$C_{var} = C_{var0} + C_{var1}v_{ds} + C_{var2}v_{ds}^2 \qquad (5.32)$$

Adding both nonlinear capacitances in (5.31) and the inverse nonlinear capacitance (5.32) compensate the total amount of nonlinear drain parasitic capacitance by tuning the varactor PMOS active nonlinear compensator.

$$C_{compensator} = C_d + C_{var} \qquad (5.33)$$

The varactor-based active PMOS distortion linearizer compensates for drain capacitance nonlinearities. Figure 5.4b illustrates that the linearity of the parasitic drain capacitance in the CMOS distributed structure can be adjusted by adding an inverse parallel varactor PMOS device to reduce the nonlinearity of the overall DA drain capacitance hence achieving highly-linear performance [97]. Figure 5.7 shows the proposed schematic of the linearized CMOS bidirectional DA.

The bidirectional properties of a conventional distributed amplifier have been compared to that of an ideal duplexer/circulator [19, 21]. The four-port distributed amplifier is inherently bidirectional because of the symmetry in its architecture. It can be excited either from port (1) or port (4). Thus it can be driven from both ends of the gate lines simultaneously. In Fig. 5.7, the proposed CMOS bidirectional distributed amplifier based tunable active duplexer has port (1) and port (4) as input ports and port (2) and port (3) as output ports. Signal power fed into port (1) emerges from port (TX2) as depicted in Fig. 5.7 and isolation is provided between port (1) and port (3).

We define the gain of the distributed amplifier as the ratio of the forward output power at port (2) to the input power at port (1) and the directivity for the distributed amplifier as the ratio of the reverse output power at port (3) to the forward output power at port (2). In dB this is given as [19, 20, 73, 107]

$$G = 10 \log \left[P^+_{out} / P^+_{in} \right] = 20 \log [S_{21}] \tag{5.34}$$

$$D = -10 \log \left[P^-_{out} / P^+_{out} \right] = -20 \log \left[S_{31} / S_{21} \right] \tag{5.35}$$

The distributed amplifier based active duplexer directivity can be improved through the tuning of S_{31} isolation over broad bandwidth. In the case of ideal distributed amplifier with no losses, S_{21} and S_{31} can be defined as [19, 20, 73, 107]

$$S_{21} = -\frac{1}{2} Z_\pi e^{-j\beta N} \left[\sum_{i=0}^{N} g_{m_i} \right] \tag{5.36}$$

$$S_{31} = -\frac{1}{2} Z_\pi \left[\sum_{i=0}^{N} g_{m_i} e^{-j\beta 2i} \right] \tag{5.37}$$

where β is the phase shift per π-section along the lines and n represents the number of devices in the amplifier. Z_π is the π-section image impedance of the drain line and g_{m_i} is the transconductance of the ith device.

From (5.34) and (5.36) the gain of the amplifier can be expressed as [19, 20, 73, 107]

$$G = 20 \log \left| \frac{1}{2} Z_\pi e^{-j\beta N} \left[\sum_{i=0}^{N} g_{m_i} \right] \right|. \tag{5.38}$$

From (5.35) and (5.37) the directivity of the amplifier is

$$D = 20 \log \frac{\left| -\frac{1}{2} Z_\pi e^{-j\beta N} \left[\sum_{i=0}^{N} g_{m_i} \right] \right|}{\left| -\frac{1}{2} Z_\pi \left[\sum_{i=0}^{N} g_{m_i} e^{-j\beta 2i} \right] \right|}. \tag{5.39}$$

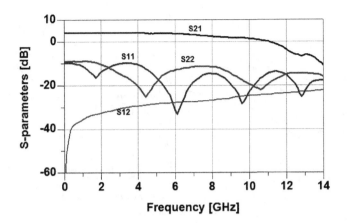

Fig. 5.8 Simulated differential S_{21} power gain, S_{11} and S_{22} return losses and S_{12} isolation for the three-stage fully-differential linearized CMOS bidirectional distributed amplifier

From (5.34) the directivity can be expressed in the form [19, 20, 73, 107]

$$D = G - 20\log\left|-\frac{1}{2}Z_\pi\left[-20\log\sum_{i=0}^{N}g_{m_i}e^{-j\beta 2i}\right]\right|. \qquad (5.40)$$

The power gain increases with the increase of number of DA gain cell stages. Amplification gain stages are connected so that output currents are combined in an additive manner at the output terminal. The gain can be increased by introducing more sections. As the RF input signal travels down the gate transmission-line, each FET transistor is excited by the traveling power wave and transfers the signals to the drain line through its transconductance. The advantages of a distributed amplifier topology are its wide bandwidth, flat gain and compact size circuit size. However, in the presence of attenuation, the gate wave signal decays as it propagates down the line. Hence, there will be a point at which the gain added by an additional device will not overcome the losses induced by the extra section in the gate and drain lines. The reason for this is that the devices added are not driven sufficiently to overcome the losses in the drain line cause of high signal attenuation.

Figure 5.8 also shows the simulated return loss S_{11}, S_{22} and S_{21} differential power gain of the three-stage bidirectional distributed amplifier peaks at 6 dB and then rolls off to a unity gain bandwidth of 11.5 GHz. The simulated input and output matching S_{11} and S_{22} are both below -10 dB indicating less of the power transferred would be reflected. The simulated isolation S_{12} performance is better than -26 dB.

Simulated IIP3 before and after linearization results for linearized CMOS bidirectional distributed amplifier are shown in Fig. 5.9. A 10 dB IIP3 improvement can be seen from Fig. 5.9.

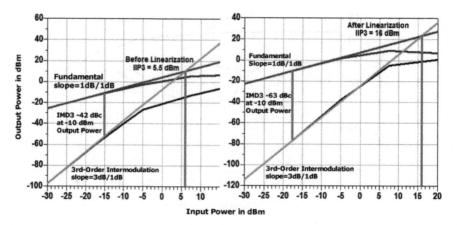

Fig. 5.9 Simulated IIP3 before and after linearization for linearized CMOS bidirectional distributed amplifier

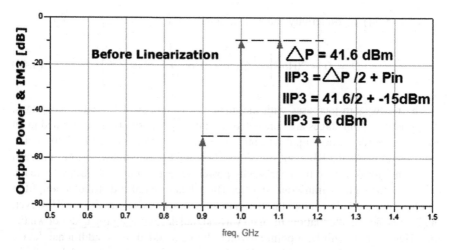

Fig. 5.10 Simulated IM3 before linearization at 1 GHz for linearized CMOS bidirectional distributed amplifier

Intermodulation distortion (IMD) nonlinearity appears as a result of applying 2 tones to the distributed bidirectional amplifier at 1 GHz with IP3 of 6 dBm. Simulated IM3 before linearization is shown in Fig. 5.10.

The degree of intermodulation distortion (IMD) nonlinearity appears less with IP3 improving to 12 dBm. Simulated IM3 after linearization at 1 GHz for linearized CMOS bidirectional distributed amplifier is shown in Fig. 5.11.

Intermodulation distortion (IMD) nonlinearity appears as a result of applying 2 tones to the distributed bidirectional amplifier at 5 GHz with IP3 of 1.8 dBm. Simulated IM3 before linearization at 5 GHz for linearized CMOS bidirectional distributed amplifier is shown in Fig. 5.12.

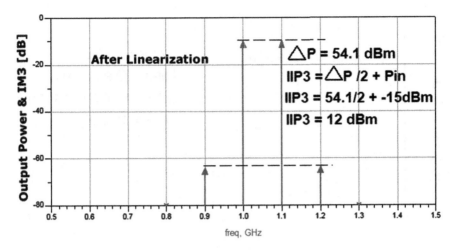

Fig. 5.11 Simulated IM3 after linearization at 1 GHz for linearized CMOS bidirectional distributed amplifier

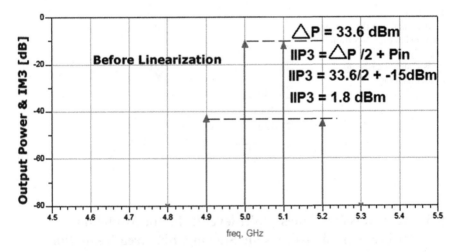

Fig. 5.12 Simulated IM3 before linearization at 5 GHz for linearized CMOS bidirectional distributed amplifier

The degree of intermodulation distortion (IMD) nonlinearity appears less with IP3 improving to 10.3 dBm. Simulated IM3 after linearization at 5 GHz linearized CMOS bidirectional distributed amplifier is shown in Fig. 5.13.

Simulated fundamental and third-order intermodulation distortion for one-stage CMOS cross-coupled cascomp transconductor is shown in Fig. 5.14.

Simulated stability factor K_f and B_{1f} characterizing the linearized bidirectional distributed amplifier stability is shown in Fig. 5.15.

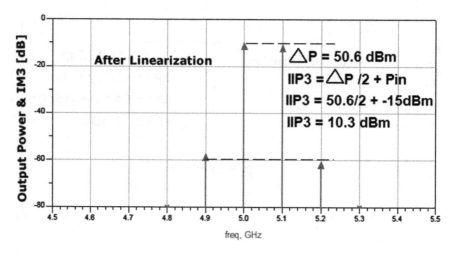

Fig. 5.13 Simulated IM3 after linearization at 5 GHz for linearized CMOS bidirectional distributed amplifier

Fig. 5.14 Simulated fundamental and third-order intermodulation distortion for one-stage CMOS cross-coupled cascomp transconductor

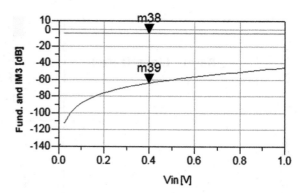

5.5 Transmission-Lines Multi-level Inductor Modeling in Transmission-Lines for Silicon Chip Area Reduction

The CMOS linearized distributed amplifier has been optimized using multi-level lumped inductor elements in order to obtain a circuit that is silicon area efficient. Further reduction in chip area is achieved using multi-level inductors [32, 108, 109]. These multi-level structures benefit from strong mutual coupling between vertically adjacent metal layers, and can generate the same inductance in less area as compared with planar inductors as shown in Fig. 5.16.

Two-port interconnect lumped model for multi-level inductor broadband equivalent circuit is shown in Fig. 5.19. Equivalent circuit model parameters are extracted from the frequency dependent quasi-static solution of the physical substrate structure and consists of ideal R, C and L components. Substrate loss for

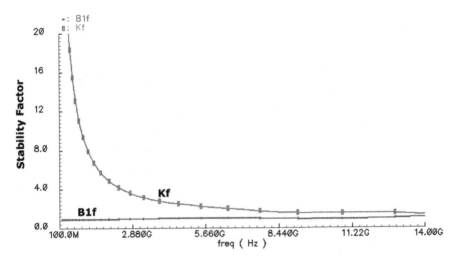

Fig. 5.15 Simulated stability factor K_f and B_{1f} characterizing linearized bidirectional distributed amplifier stability

Fig. 5.16 HFSS linearized CMOS bidirectional distributed amplifier transmission-lines multi-level inductor modeling RF CMOS 0.13 μm. The HFSS modeled inductor has an outer diameter of 91 μm and spacing of 5 μm with width of 10 μm

interconnect is modeled by the resistor R_{sub} and capacitor network that consists of C_{ox} and C_{sub} representing the oxide layer parasitic capacitances between the conductors and the bulk substrate [75, 97, 107, 109] (Fig. 5.19).

Accurate modeling at microwave frequencies requires electromagnetic simulations using ADS Momentum or HFSS EM engines as shown in Fig. 5.18. A 0.13 μm CMOS silicon substrate was constructed in both HFSS and ADS Momentum, based on the available data from process foundry Fig. 5.17. The IBM 0.13 μm CMOS process offers three thick RF metal layers suitable for high-Q inductors and the top metal layer (MA) is composed of aluminum. In order to reduce the loss and improve the quality factor (Q), the metallization layer with the lowest loss is chosen

Fig. 5.17 ADS linearized
CMOS bidirectional
distributed amplifier
transmission-lines multi-level
inductor modeling RF CMOS
0.13 μm

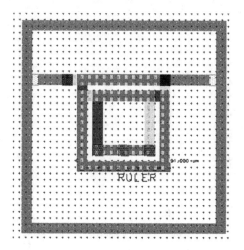

for the design. Inductor Q as shown in Fig. 5.18 is frequency dependent. It is clear that inductor Q is strongly affected by the metal thickness (which related to metal loss) and substrate resistivity.

Stacked inductors implemented in two or three metal layers were designed with inductance values 900 and 450 pH. The gate transmission-line 450 pH inductor has an outer diameters of 70 μm and spacing of 5 μm and width of 9 μm and number of turns n of 1.5. The drain transmission-line 900 pH inductor has an outer diameter of 91 μm and spacing of 5 μm and width of 10 μm and number of turns n of 1.75. Both stacked inductors were implemented on MA and E1 top metal layers which corresponds to the eight and seventh top metal layers. The gate transmission-line m-derived 280 pH inductor has an outer diameter of 59 μm and spacing of 5 μm and width of 9 μm and number of turns n of 1.5. The drain transmission-line m-derived 230 pH inductor has an outer diameter of 55 μm and spacing of 5 μm and width of 9 μm and number of turns n of 1.5.

As with planar inductors, reducing area over substrate is paramount in increasing the resonance frequency (SRF) of stacked inductors [15, 75, 97, 107]. The nearly 50° reduction in total area with more metal layers yields higher SRF, even though the bottom metal layer is slightly closer to the substrate. EM simulations in HFSS and ADS were performed on the gate and drain transmission-lines multi-level inductors. The EM simulation results of Q and inductance L in RF CMOS 0.13 μm are shown in Fig. 5.18. The ADS EM calculations were performed using the methods of moments where as in HFSS the 3D full-wave EM calculations were performed using the finite element method. The EM simulations results for both calculations are shown in Fig. 5.18.

CMOS processes have low resistive substrates which cause significant losses which makes design of high Q inductors challenging. At low frequencies, metal losses are mainly determined by the sheet resistance of the process layers used to create the device. However, at high frequencies, skin effect, proximity effects and

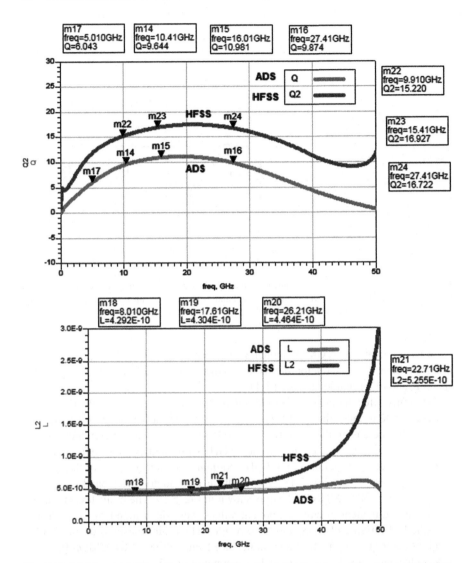

Fig. 5.18 HFSS and ADS linearized CMOS bidirectional distributed amplifier gate transmission-lines multi-level inductor modeling Q and L Modeling RF CMOS 0.13 μm. The ADS EM calculations were performed using the methods of moments where as in HFSS the 3D full-wave EM calculations were performed using the finite element method

current crowding have a major impact on the loss mechanism [27, 29, 75, 97, 107]. The skin effect drives the ac current toward the surface of the conductor. Skin effect increases the ac resistance of the conductor leading to lower Q [110–113].

Inductor performance is strongly affected by the metal loss and substrate loss. Increasing metal thickness to reduce metal loss could significantly improve inductor

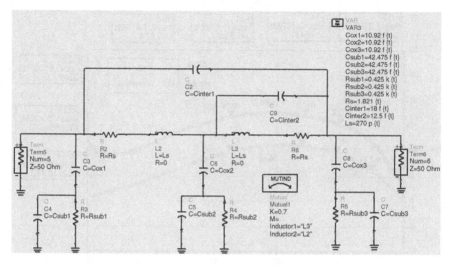

Fig. 5.19 ADS transmission-lines multi-level inductor broadband equivalent circuit model in RF CMOS 0.13 μm. The multi-level inductor broadband equivalent circuit component values were determined by curve-fitting with the EM HFSS and ADS simulated results

Q for RF applications. A CMOS inductor's performance is also affected by the oxide thickness beneath the inductor and its lateral dimensions, such as metal strip width, spacing, and outer diameter [65, 114, 115].

Proximity effect metal loss degrades the performance of on-chip inductor at high frequency due to the influence of the magnetic field created by a nearby conductor and thereby increasing the effective series resistance. The two-level inductor has a peak of approximately 15 and it peaks at 25 GHz with a self-resonance frequency of 50 GHz. Effective inductance of 450 pH and quality factor Q of 13 for the stacked inductor are shown in Fig. 5.18. The multi-level inductor broadband equivalent circuit model in RF CMOS 0.13 μm is shown in Fig. 5.19. The multi-level inductor broadband equivalent circuit component values were determined by curve-fitting with the EM HFSS and ADS simulated results. The drain transmission-line multi-level inductor quality factor Q and inductance L modeling for RF CMOS 0.13 μm with inductance of 910 pH and Q of 15 is shown in Fig. 5.20.

5.6 DA Based Duplexer with Integrated Antenna on Silicon Replacing DA M-Derived Matching Network

An on-chip differential loop antenna is suitable for an antenna/inductor design as it is inherently inductive in nature and delivers a broad radiation pattern [13, 16]. This antenna/inductor structure is not operating at resonance and is integrated on a lossy

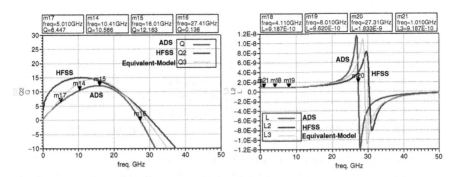

Fig. 5.20 HFSS and ADS drain transmission-line multi-level inductor Q and L modeling for RF CMOS 0.13 μm

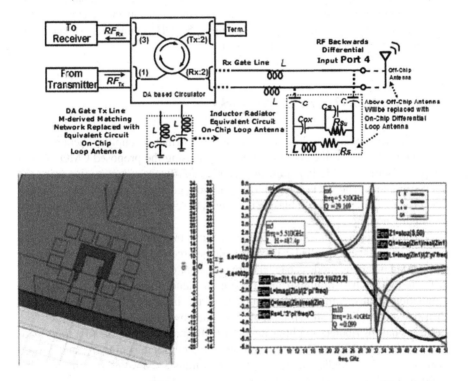

Fig. 5.21 ADS simulated inductance and quality factor curve fitting for on-chip differential loop antenna inductor radiator and its equivalent circuit for L = 0.49 nH with Q = 29 at 5.5 GHz

silicon substrate [16]. The inductor is optimized to radiate efficiently and therefore serves as an on-chip radiating antenna as well.

The proposed CMOS bidirectional DA based tunable active duplexer connects the radiating inductor element to its receive path input port as can be seen in Fig. 5.21. The on-chip differential loop antenna replaces the gate line m-derived image impedance transforming section on receive gate line input port. The signal attenuation along the gate line can be overcome by amplifying the signal with the DA gain cells.

The on-chip differential loop antenna provides the required inductance to replace the bidirectional DA based active duplexer m-derived inductor component. The on-chip differential loop antenna with dimension $280 \times 220\,\mu m$ designed in HFSS, exhibits a simulated inductance of $L = 0.49\,nH$ replacing the required m-derived matching network inductance performance and has a quality factor of $Q = 29.1$ at 5.5 GHz as can be seen in Figs. 5.21 and 5.22.

Figure 5.22 shows an HFSS model for a stand alone on-chip differential loop antenna with matched testing loop antenna model. Each antenna is fed differentially through a wave port. The S_{21} power is extracted for both the on-chip loop antenna and the matched testing loop antenna. Simulation results in Fig. 5.22 show that the co-design DA based tunable active duplexer with on-chip loop antenna has 6 dB S_{21} power gain added improvement from DC up to 5.2 GHz compared to stand alone on-chip loop antenna S_{21} power performance.

5.6.1 Varactor-Tuned LC Networks

The proposed linearized CMOS bidirectional distributed amplifier achieves linearization on two parameters, transconductance using the proposed CMOS cross-coupled cascomp and drain nonlinear parasitic capacitance compensation using the MOS varactors. The MOS capacitor is gate voltage dependent, so when using a MOS capacitor depending on biasing condition, the capacitance will be small, lossy, and highly nonlinear.

The test schematic showing the testbench when the MOSCAP varactor (variable reactors or voltage controlled capacitors) with source and drain connected together to act as an inversion mode capacitance by varying the voltage is shown in Fig. 5.23 with the MOSCAP minimum and maximum value of the varactor capacitance simulation results. Simulated Performance of the linearized CMOS bidirectional distributed amplifier is shown in Table 5.1. The power gain S_{21} of the three-stage bidirectional distributed amplifier peaks at 6 dB and then rolls off to a unity gain bandwidth of 11.5 GHz. The simulated input and output matching S_{11} and S_{22} are both below -10 dB indicating less of the power transferred would be reflected. The simulated isolation S_{12} performance is better than -26 dB.

5.7 Chapter Summary

In this chapter, a fully-integrated fully-differential linearized CMOS distributed bidirectional amplifier that achieves large IMD3 distortion reduction over broadband frequency range for both RF paths was demonstrated. The drain and gate

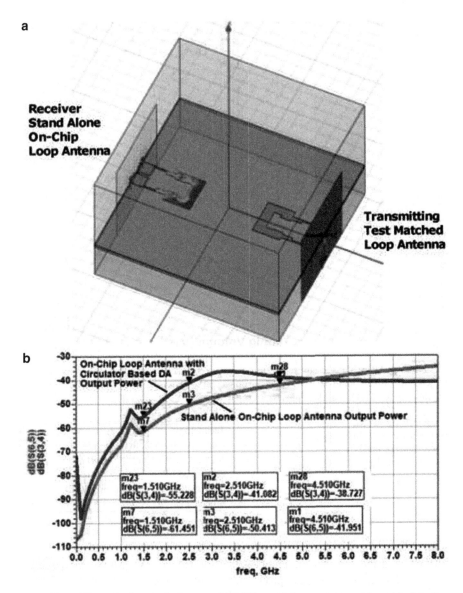

Fig. 5.22 HFSS model for stand alone on-chip differential loop antenna with matched testing antenna model. (**b**) Co-design of DA based active duplexer with on-chip loop antenna has 6 dB S_{21} power gain added improvement up to 5.2 GHz compared to stand alone on-chip loop antenna S_{21} power performance. The on-chip differential loop antenna with dimension $280 \times 220\,\mu m$ designed in HFSS

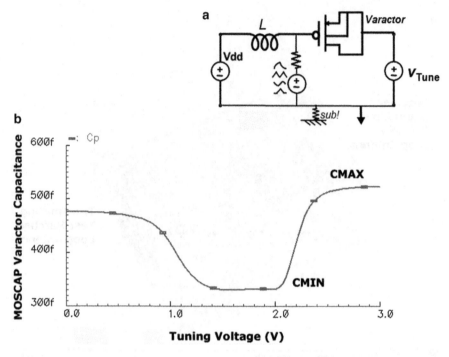

Fig. 5.23 (**a**) MOSCAP varactor test setup schematic with source and drain connected together to generate a highly non-linear capacitance-voltage curve behavior (**b**) C-V characteristic of the minimum and maximum MOSCAP varactor capacitance

Table 5.1 Simulated performance of the linearized CMOS bidirectional distributed amplifier

Technology	RF CMOS 0.13 μm
Unity gain bandwidth	11.5 GHz
S_{21} peak power gain	6 dB
Linearized IIP3	>10 dB
S_{12} isolation	<−26 dB
Silicon area	1.887 × 0.795 mm

transmission-lines were stagger-compensated. Reducing the DA IM3 distortion by mismatching the gate and drain LC delay-line ladders. A CMOS cross-coupled compensator transconductor is proposed to enhance the linearity of the DA gain cell with a varactor-based active post nonlinear drain capacitance compensator for wider linearization bandwidth.

Chapter 6
Linearized CMOS Distributed Bidirectional Amplifier Silicon Chip Implementation

6.1 Introduction

This chapter presents several practical layout guidelines of the fully-integrated fully-differential linearized distributed bidirectional amplifier implemented in IBM CMOS RF 0.13 μm silicon process. The total chip silicon area is 1.5 mm^2 including testing pads. The circuit elements forming the linearized CMOS bidirectional distributed amplifier are discussed in terms of their physical arrangement and layout.

6.2 Linearized CMOS Bidirectional Distributed Amplifier High Frequency Layout Considerations

CMOS technologies offer the capability of integrating both baseband and RF front-end transceiver components on a single chip allowing low cost implementation [116]. However, CMOS has inherent technology limitations such as low current drive, lossy substrate, low breakdown voltage and poor transconductance [32–34]. These technology drawbacks have negative impact on implementing fully-integrated power amplifier using CMOS process. For this reason, a fully differential topology has been adopted since differential power amplifier have several advantages over single-ended ones. For instance, the voltage doubling effect lowers the burden of low breakdown voltage limit. Also, the virtual ground of source prevents gain reduction from source inductance degeneration and stability can be easily achieved [32–34].

The design of the proposed linearized bidirectional DA has been fully-integrated in 0.13 μm RF CMOS technology. Proper layout techniques are used such as maintaining device matching and use of symmetry in circuit layout design. The use of balanced modular layout design results in balanced current distribution and supply routing [32–34]. With interdigitating wide transistors, with multi-gate finger layout, the gate resistance of the polysilicon becomes smaller. Grounded "guard ring"

Z. El-Khatib et al., *Distributed CMOS Bidirectional Amplifiers: Broadbanding and Linearization Techniques*, Analog Circuits and Signal Processing, DOI 10.1007/978-1-4614-0272-5_6, © Springer Science+Business Media New York 2012

are added surrounding sensitive active devices to improve device isolation and acts as an effective shield to noise and crosstalk. ESD were added to the gates in order to protect the chip from antenna effects [32–34].

Technology process parameters such as substrate resistivity, number of metal layers, distance between metal layers, and metal layer thickness are all set by 0.13 μm CMOS technology [117]. The 0.13 μm CMOS top most metal layer has lower sheet resistance and suffers less loss. The top metal layer has the largest thickness which helps in improving component quality factor Q and it is commonly utilized for routing critical high frequency signals. CMOS 0.13 μm technology offers eight metal layer (M1–M8) [117] including three thin metal (M1, M2, and M3) layers, two thick metal (MQ and MG) layers and three RF (LY, E1, and MA) layers. Modeling the substrate definition includes the number of layers, position of each layer and composition of each layer.

The drain transmission line signal conductor dimensions are made wide in order to reduce the resistivity on the drain transmission lines which greatly affects the high-frequency performance of the device. Gate tie-downs where added to long metal traces in order to cancel antenna effect preventing the collection of charges that may destroy the device gates. ESD double diode protection were added to I/O pads and multiple contacts are placed where needed to ensure good connections [32–34].

6.3   Silicon CMOS RF Multi-level Inductors Implementation

On-chip inductors play a crucial role in radio frequency integrated circuits. Several factors degrade the performance of on-chip passive components at high frequencies such as skin effect, proximity effect, electric field penetration into substrate, and substrate eddy current losses [20]. The main parameters that characterize an inductor are its self-inductance value, L, its quality factor, Q, and its resonant frequency f_{res} [75].

A good understanding of loss mechanisms is essential in designing silicon RF CMOS on-chip inductors. Loss mechanisms include metal losses, substrate losses and high frequency losses [20]. The main source of the loss of the on-chip inductor is the conductive silicon substrate. At lower frequencies, metal losses are mainly determined by the sheet resistance of the process layers used to create the device [39]. At higher frequencies, skin effect, proximity effects and current crowding have a major impact on loss mechanism [118].

The skin effect drives the ac current toward the surface of the conductor [118]. Skin effect increases the ac resistance of the conductor leading to lower Q. Eddy current substrate losses are caused by the magnetic field of the conductor inducing a current in the substrate [118].

Strong mutual coupling in multi-layer inductors allows the generation of the same inductance in less area as compared to planar inductors [32]. In distributed broadband amplifiers, a large number of inductors are necessary, leaving a small

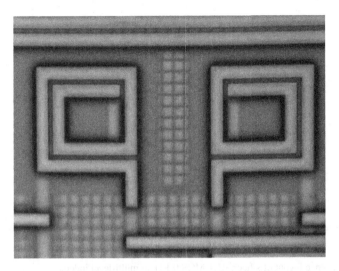

Fig. 6.1 Micrograph of silicon RF CMOS 0.13 μm multi-level drain line inductors with outer diameter of 91 μm and spacing of 5 μm and width of 10 μm and number of turns n of 1.75

silicon area for each [17]. A good candidate for this purpose is the stacked structure shown in Fig. 6.1, where two planar spirals are placed in series each on different metal layer.

Multi-level inductors can be implemented by connecting two or more layers in series [108]. This will let the same silicon area to be realized with an increased inductance value. Since the current flows through both metal layers in the same direction, the magnetic flux lines in the same direction result in higher mutual inductance. The flux through the two windings will strengthen one another and the total inductance of a stacked spiral consists of the sum of the two planar spiral inductances plus their mutual inductance [108]. One drawback with this approach is the lower self-resonant frequency resulting from the large feed through capacitance between the layers [111].

Square inductors are generated easily with ASITIC [119]. The physical parameters have to be determined, including number of turns N, conductor width W, the edge-to-edge spacing between adjacent turns, S, and the outer diameter, OD. The geometrical parameters were chosen to realize an inductance value while minimizing parasitic capacitance and area. ASITIC generates the inductor pi model parameters of spiral inductors through electromagnetic field solutions at a given frequency [119]. The layout is then imported to Cadence and ADS simulation environments.

Separation between on-chip spiral inductors should be large enough to minimize magnetic coupling effects between them as shown in Figs. 6.2 and 6.1 [120]. On-chip separation between inductors involves spacings that is a function of the layout size of the inductor [29]. Surrounding the CMOS 0.13 μm multi-level inductors is a metallization ground shield rings as shown in Fig. 6.2 which acts as the path

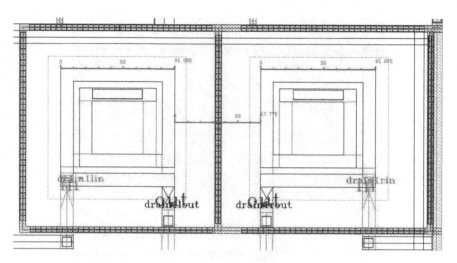

Fig. 6.2 Cadence layout of silicon RF CMOS 0.13 μm multi-level inductor

for the return current in the structure and to reduce the substrate loss. The guard ring consists of P + or N + diffusion regions connected to ground. The guard ring prevents eddy currents from being induced in the substrate. The staggered transmission-line gate (450 pH) and drain (900 pH) on-chip inductors are modeled using both Ansoft's High Frequency Structure Simulator (HFSS) and Momentum ADS.

6.4 CMOS Bidirectional Distributed Amplifier Cross-Coupled Compensator Gain Cell Layout

Layout techniques such as interdigitated transistor layouts, symmetry for transmission-line differential paths, multiple finger transistor layout to minimize gate resistance [32–34]. Guard rings were placed surrounding the devices and inductors to reduce noise. All these RF layout techniques adopted to enhance the CMOS 0.13 μm linearized bidirectional distributed amplifier performance as shown in Fig. 6.3.

Minimizing the MOSFET gate resistance was achieved by properly sizing the finger width used during the layout of the transistor to $W_{finger} = 4$ μm. Another important consideration in the linearized CMOS bidirectional distributed amplifier layout is the series gate resistance of the MOS transistor since it reduces gain significantly at higher frequencies. To prevent this effect, the gate poly was split into many fingers as depicted in Fig. 6.3 in order to keep the total resistance down as shown in Fig. 6.4.

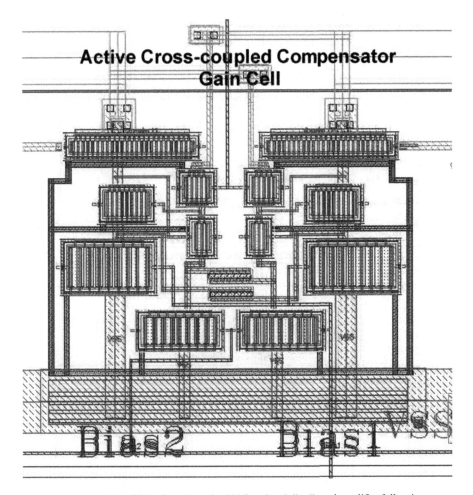

Fig. 6.3 Silicon RF CMOS 0.13 μm linearized bidirectional distributed amplifier full active cross-coupled compensator gain cell with enhanced tunability

6.5   Linearized CMOS Bidirectional Distributed Amplifier Full Layout

Most of the interconnections are done at the top metal eight layer (MA) which has the highest conductivity Fig. 6.5. The first metal layer was used as a ground plane (Fig. 6.6). The layout of the gain cell blocks are done symmetrically as can be seen in Fig. 6.6. The width of metal interconnections was chosen according to the amount of current flowing through them. In order to meet pattern density requirements for the layout, filling cells are placed in the blank space surrounding the gain cells as shown in Figs. 6.6 and 6.7. Eight pin probing pads are used in Cadence layout design of the linearized bidirectional amplifier.

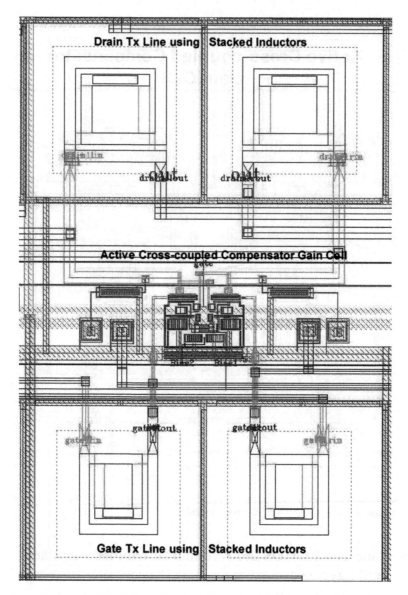

Fig. 6.4 Silicon RF CMOS 0.13 μm linearized bidirectional distributed amplifier cross-coupled compensator gain cell with multi-level inductors

As can be seen in the layout in Figs. 6.6 and 6.7, the chip area is dominated by the passive on-chip inductors components. The micrograph of the proposed fully-integrated fully-differential silicon RF CMOS 0.13 μm linearized bidirectional distributed amplifier co-designed with the on-chip antenna is shown in Fig. 6.7. The total chip silicon area is 1.887×0.795 mm^2 including testing pads.

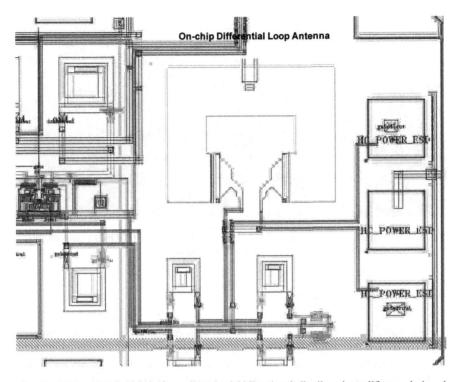

Fig. 6.5 Silicon RF CMOS $0.13\,\mu m$ linearized bidirectional distributed amplifier co-designed with on-chip antenna. The on-chip differential loop antenna with dimension $280 \times 220\,\mu m$ designed in HFSS

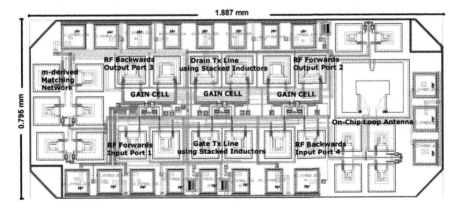

Fig. 6.6 Fully-integrated silicon RF CMOS $0.13\,\mu m$ linearized bidirectional distributed amplifier layout co-designed with on-chip antenna with multi-level inductors. The total chip silicon area is $1.5\,mm^2$ including testing pads

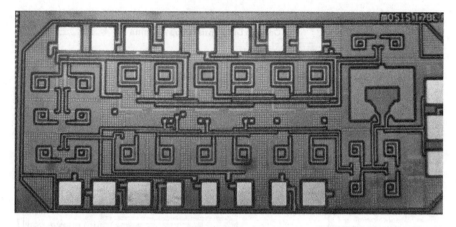

Fig. 6.7 Micrograph of fully-integrated fully-differential silicon RF CMOS $0.13\,\mu m$ linearized bidirectional distributed amplifier co-designed with on-chip antenna

6.6 Chapter Summary

In this chapter, high-frequency layout design techniques used in implementing the fully-integrated linearized distributed bidirectional amplifier in silicon RF CMOS $0.13\,\mu m$ process have been discussed. The circuit elements forming the linearized CMOS distributed bidirectional amplifier were presented in terms of their physical arrangement and layout.

Chapter 7
Linearized CMOS Distributed Bidirectional Amplifier Experimental Setups and Chip Measurement Results

7.1 Introduction

This chapter presents the experimental test setups for the fully-integrated linearized CMOS RF 0.13 μm bidirectional distributed amplifier. The chip measurements that are carried out include small-signal s-parameters such as gain and return loss, harmonics power measurements such as 1-dB compression, IMD (intermodulation distortion) and noise figure measurements. All RF measurements are performed with on-wafer probing.

7.2 High-Frequency On-Wafer Measurement System

All RF measurements were carried out with on-wafer probing. On-wafer probing as shown in Figs. 7.1 and 7.3 was performed on an analytical probe station eliminating all of the problems inherent in packaging removing any uncertainty about bond wire inductances. The measurement test bench setup included network analyzers, high-frequency wafer probes and programmable DC source units.

Automated chip measurement characterization are carried out with test instruments connected using GPIB. To minimize test time, matlab scripts were compiled to control the instruments. The results are instantly graphed to view acquired measurement data from test equipment through the GPIB interface.

The testing instruments include an RF signal generator, a power meter and Agilent RF signal generator as shown in the Fig. 7.1. The testing instruments also includes an Agilent RF spectrum analyzer for characterizing frequency content, a vector network analyzer, modulation analyzers for characterizing modulated content in the presence of undesired blocking signals such as those from simulated cellular base stations, other cellphones, and broadcast radio stations Fig. 7.2. The measurements were conducted utilizing 8-pin high-frequency wafer probes.

Z. El-Khatib et al., *Distributed CMOS Bidirectional Amplifiers: Broadbanding and Linearization Techniques*, Analog Circuits and Signal Processing, DOI 10.1007/978-1-4614-0272-5_7, © Springer Science+Business Media New York 2012

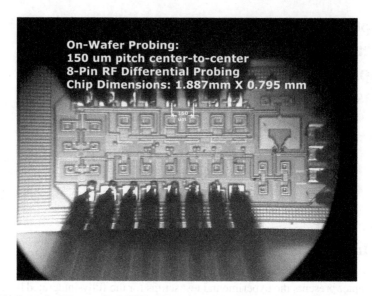

Fig. 7.1 On-wafer RF automated GPIB chip testing of the 0.13 μm RF CMOS linearized CMOS distributed bidirectional amplifier

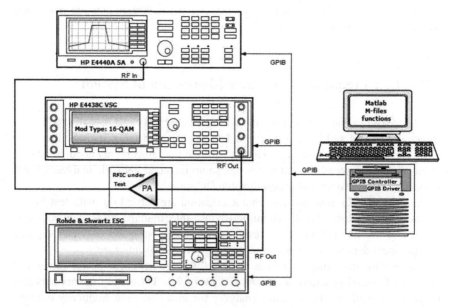

Fig. 7.2 RF automated GPIB testing of the 0.13 μm RF CMOS three-stage fully-differential linearized CMOS distributed bidirectional amplifier with high-frequency RF probes

7.3   S-Parameter and Harmonics Power Measurements

The S-parameter test setup is shown in Fig. 7.3 and includes DC-supplies, 8-pin high
frequency probes and a vector network analyzer. The S-parameter measurements
were carried out using the Anritsu 37347C Vector Network Analyzer and multi-
contact wedge 8-pin high frequency differential probes with two RF needles up to
40 GHz. The network analyzer is capable of accurately extracting the S-parameters
of the device up to 20 GHz. The Anritsu 37347C was calibrated up to the probes with
the Open-Short-Load-Through calibration method [22, 67, 118, 121]. A full two-
port calibration from a start frequency of 50 MHz to a stop frequency of 20 GHz
are performed. The device under test was connected as shown in Fig. 7.3. The
measured data collected using network analyzer in the form of S-parameter relating
the electromagnetic waves scattered from the device under test to those EM waves
incident upon the vector network analyzer.

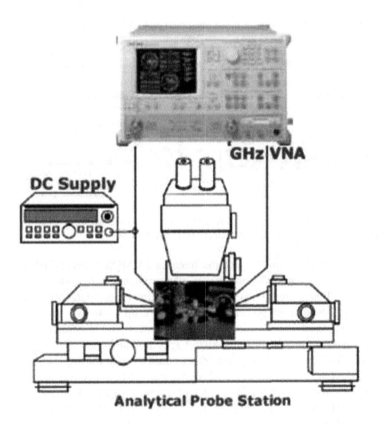

Fig. 7.3 Test setup for S-parameter measurements for the linearized CMOS distributed bidirectional amplifier

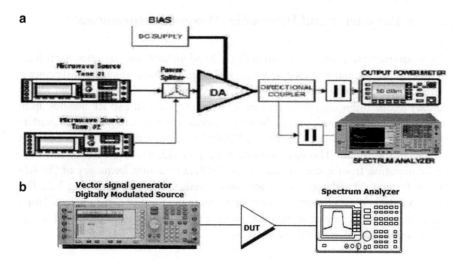

Fig. 7.4 (**a**) Two-tone test setup for measuring power and intermodulation distortion. (**b**) Vector digitally modulated signals generator measurement setup

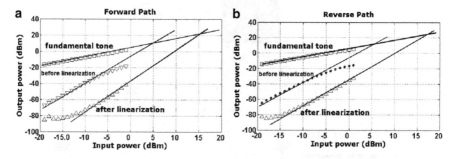

Fig. 7.5 IIP3 Comparison of measured reverse path output power versus input power with and without linearization for both (**a**) forward and (**b**) reverse RF paths

To characterize the linearity of the linearized CMOS bidirectional distributed amplifier, one-tone and two-tone power measurements were conducted for characterizing 1-dB compression point P_{1dB} and third-order interception point (IP3) respectively. The output power measurement setup is shown in Fig. 7.4. IIP3 measurements used two Rohde and Shwartz tone signal generators and a high-frequency power combiner with the 8975A spectrum analyzer.

An external 12 GHz power combiner was used to combine the tones from signal generators. At the output stage, the RF signal was fed to a spectrum analyzer which provided a load resistance of $50\,\Omega$. On-wafer two-tone IMD3 measurements was performed. The 8975A spectrum analyzer was used for single-tone compression measurements as well.

Two-tone power measurements were conducted to characterize the third-order input interception point (IIP3) of the linearized CMOS bidirectional DA. IIP3 comparison of measured output power versus input power with and without linearization is shown in Fig. 7.5. A measured IIP3 improvement of 10 dB at 5 GHz

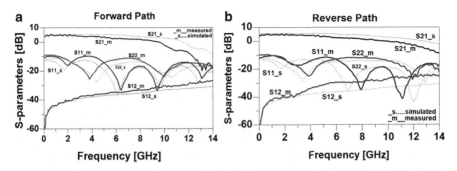

Fig. 7.6 Comparison of simulation and measured differential S-parameters for both (**a**) forward and (**b**) reverse RF paths

with 100 MHz spacing is achieved for both forward and reverse RF paths. The CMOS linearized bidirectional amplifier achieves a highly linear output power of 18.5 dBm.

The small-signal S-parameters measurements were carried out using an Anritsu 37347C vector network analyzer with proper calibration techniques. The measured RF forward path and reverse path S-parameters for the linearized CMOS bidirectional DA achieves a measured peak gain of 5 dB and unity gain bandwidth of 9.5 GHz as shown in Fig. 7.6a, b for both forward and reverse RF paths. The measured input and output return losses are better then −10 dB over the frequency range with better than −26 dB isolation.

The IM3 distortion reduction measurement of the linearized bidirectional amplifier was examined at four different two-tones frequencies at (1, 3, 5 and 5.9 GHz) with 100 MHz spacing using an Agilent E4440A spectrum analyzer. A comparison of the measured output spectra with and without linearization for these frequencies is shown in Figs. 7.7–7.10. Intermodulation distortion (IMD) nonlinearity appears as a result of applying two tones to the distributed bidirectional amplifier. The measured IM3 distortion reduction is 20 dB for both forward and reverse RF paths. The intermodulation distortion (IMD) nonlinearity improvement match the simulation results. A comparison of the measured IM3 distortion reduction with and without linearization over broadband frequency of operation for both RF paths is shown in Fig. 7.11. The linearization over broadband frequency measurement characterization were carried out with both RF signal generators and the Agilent E4440A spectrum analyzer test instrument connected using GPIB. The intermodulation distortion (IMD) nonlinearity results are instantly graphed to view acquired measurement data from spectrum analyzer test equipment.

Unmodulated carrier signals have been used as stimuli for distortion measurements. However, digitally-modulated stimulus signals are used as well as they provide more realistic measurement results [122]. For a digitally modulated carrier, distortion produces spectral regrowth. Nonlinear spectral analysis with digitally modulated signal as input was carried out [122]. To demonstrate the

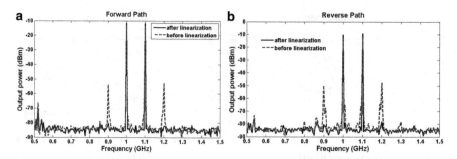

Fig. 7.7 Comparison of measured output spectra with and without linearization for two-tones frequencies at 1 GHz with 100 MHz spacing for both (**a**) forward and (**b**) reverse RF paths

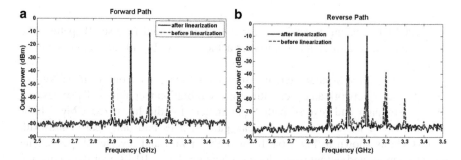

Fig. 7.8 Comparison of measured output spectra with and without linearization for two-tones frequencies at 3 GHz with 100 MHz spacing for both (**a**) forward and (**b**) reverse RF paths

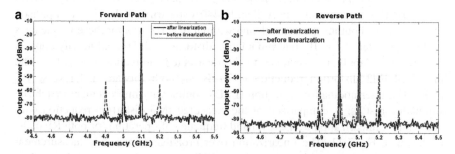

Fig. 7.9 Comparison of measured reverse path output spectra with and without linearization for two-tones frequencies at 5 GHz with 100 MHz spacing for both (**a**) forward and (**b**) reverse RF paths

linearization capability of the bidirectional amplifier, it was measured using an Agilent E4438C ESG vector signal generating communication-specific test signals for testing wireless communication systems such as Π/4-differential QPSK (DQPSK) modulated signals [122]. These modulation schemes are commonly used in many commercial wireless systems on the market today due to their spectral efficiency and

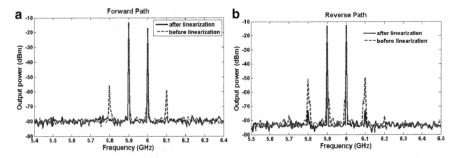

Fig. 7.10 Comparison of measured output spectra with and without linearization for two-tones frequencies at 5.9 GHz with 100 MHz spacing for both (**a**) forward and (**b**) reverse RF paths

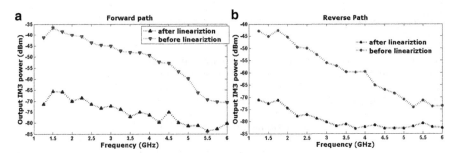

Fig. 7.11 Comparison of the measured IM3 distortion reduction with and without linearization over broadband frequency of operation for both (**a**) forward and (**b**) reverse RF paths

simple modulator/demodulator construction. A performance test was done using a narrowband Π/4-DQPSK signal. The signal used for this test is a standard Π/4-DQPSK signal at 125 ksymbols/s (250 kbits/s) and was pulse shaped using a square root Nyquist 40 MHz filter using a roll-off factor $\beta = 0.35$.

Measured normalized output power spectral density PSD with and without linearization at 2 GHz for a standard Π/4-DQPSK digitally modulated carrier signal is shown in Fig. 7.12. We can see that the nonlinearity of the amplifier has caused the output signal to be spread in frequency and the third order nonlinearity has caused the first set of shoulders on the output signal, which are around 45 dB down from the signal peak. This output signal is unacceptable from a system perspective, as the distortion spills out of the channel used by the signal and into nearby channels that are occupied by signals from other users. This is known as adjacent channel interference, and reduces the spectral efficiency of the system. Spectral regrowth due to amplifier nonlinearities are fully compensated for, resulting in a 20 dB improvement in adjacent channel interference relative to the output signal.

To further demonstrate the performance of the linearized bidirectional DA, using an Agilent E4438C ESG vector signal generator a custom multi-tone waveform test was generated using a ten-tones 5 MHz bandwidth each [122]. Measured normalized

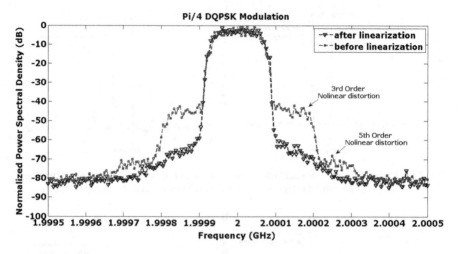

Fig. 7.12 Measured output PSD with and without linearization at 2 GHz for Π/4-DQPSK digitally modulated carrier signal using a square root Nyquist filter using a roll-off factor $\beta = 0.35$. Spectral regrowth due to amplifier non-linearities are fully compensated for, resulting in nearly 20 dB improvement in ACI relative to the output signal

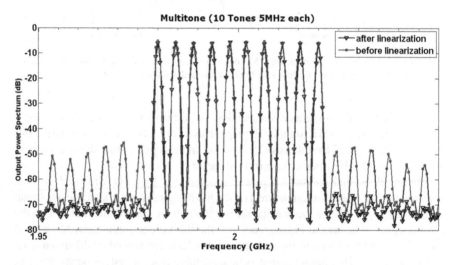

Fig. 7.13 Measured output PSD with and without linearization at 2 GHz for ten-tones 5 MHz bandwidth each multi-tone modulated signal

output PSD with and without linearization at 2 GHz for Multi-tone modulated signal is shown in Fig. 7.13. A 20 dB of improvement is achieved through linearization.

QAM is another common modulation technique used in modern communications systems due to its bandwidth efficiency. The measured normalized output PSD with and without linearization at 2 GHz for 4-QAM digitally modulated carrier signal is shown in Fig. 7.14.

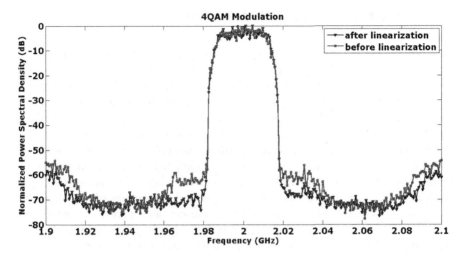

Fig. 7.14 Measured output PSD with and without linearization at 2 GHz for 4-QAM digitally modulated carrier signal

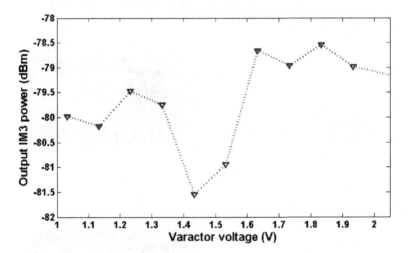

Fig. 7.15 Comparison of the IM3 distortion reduction under different varactor bias voltage adjustment from 1 to 1.8 V there exist an optimal bias at which IM3 suppression is formed at 1.45 V

The varactor-based active post distortion enables one to tune the IM3 reduction with ease and precision. It provides distortion cancellation tuning at various combination of varactor voltages. The varactor bias voltage is adjusted from 1 to 1.8 V. An optimum bias condition is observed for IM3 distortion suppression which is formed at 1.45 V as can be seen in Fig. 7.15.

7.4　Noise Figure Setup and Measurement

The noise figure test setup includes noise figure analyzer N8975A, high-frequency noise source, DC-supplies, device under test, RF 8-pin wafer probes as shown in Fig. 7.16. A noise figure measurement of the linearized CMOS bidirectional distributed amplifier was performed. After calibration, the measurements indicate that the noise figure of the amplifier shows a minimum of 7.8 dB at around 1 GHz and is below 9 dB for frequencies up to 9.5 GHz as shown in Fig. 7.17.

A summary comparing previous state-of-the-art published measured linearized DAs is presented in Table 7.1. Table 7.1 highlights the proposed linearized bidirectional DA large 20 dB IMD3 reduction over broadband frequency range for both RF paths with the least power consumption and minimum silicon chip area compared to other published resutls. Table 7.1 presents all the previously published Linearized DA's to the author's knowledge. It shows that the proposed linearized DA has the lowest power consumption and minimum silicon chip area in comparison to all the previously published linearized DAs. The proposed linearized DA is implemented with three-stage gain cells to optimize for silicon chip area of 1.5 mm^2. The proposed linearized bidirectional DA power consumption is 128 mW.

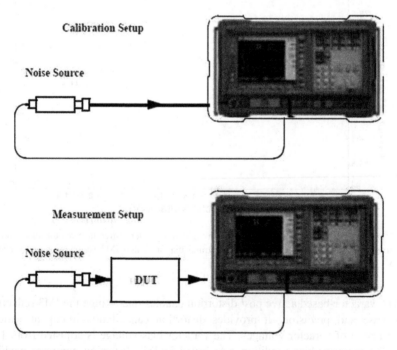

Fig. 7.16 Noise figure calibration and measurement setup for the linearized CMOS distributed bidirectional amplifier

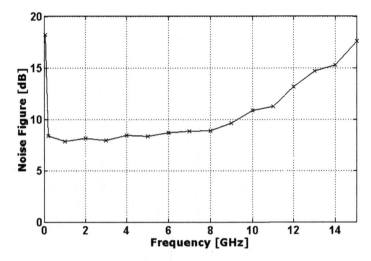

Fig. 7.17 Measured noise figure for the linearized CMOS distributed bidirectional amplifier

Several linearized DAs have been published [10–123] and [65, 66] as shown in Table 7.1, however most of the published DA linearization methods reported do not provide large IM3 distortion reduction. Since they involve system-level linearization with bulky discrete components which is not suited for fully-integrated circuit miniaturization. Due to the discrete component performance variation with frequency, they also suffered from limited linearization over broad bandwidth. Other DA linearization techniques have narrow linearized bandwidth at lower frequencies [13, 15] and apply only DC-based linearization techniques.

A CMOS DA based multi-tanh linearization technique is also reported in [26]. However the CMOS DA based multi-tanh linearization technique offered a limited 5 dB IM3 distortion reduction [26]. Another linearized DA that has been published is a differential DA with circuit-level feedforward linearization technique [27]. It operated over a wide band from 0.1 to 12 GHz, however only simulation results were presented [27]. Lau and Chan proposed a linearized DA that achieved a 10 dB IM3 reduction over a limited 2.3 GHz bandwidth range [23]. Recently, Lu and Pham [28, 29] proposed a multi-gated transistor (MGTR) topology based CMOS linearized DA. The MGTR-based linearized distributed amplifier operated over a limited bandwidth of 4 GHz range and had only a 11 dB IM3 reduction. Comparing the proposed CMOS linearized DA in this work [30] to other published ones, the proposed linearized CMOS DA offers a 20 dB IM3 distortion reduction with 9.5 GHz operational bandwidth and with the least power consumption [30].

Measurement setup of the 0.13 μm RF CMOS three-stage fully-differential linearized CMOS bidirectional distributed amplifier on analytical probe station is shown in Fig. 7.18. The testing instruments include an RF signal generator, a power meter and Agilent RF signal generator and programmable DC source units as shown

Table 7.1 Comparison with state-of-the-art previously published measured linearized distributed amplifiers highlighting the large 20 dB IMD3 reduction over broadband frequency range for both RF paths with least power consumption and minimum silicon chip area for the proposed bidirectional fully-differential fully-integrated solution

Design Ref.	Technology process	Power gain (dB)	Operational bandwidth (GHz)	IMD3 reduction (dB)	IIP3 (dBm)	Power consump. (mW)	Linearization technique	Circuit topology chip area mm^2	Differential or single ended
[17]	GaAs HEMT	NA	10	15	NA	NA	Feedforward	Discrete components	Single ended
[10]	SiGe BJT	14	2.2	12	NA	NA	Parallel diode	Discrete components	Single ended
[23]	SiGe BJT	13	2.3	10	10	180	Self biased	Discrete components	Single ended
[24]	GaAs MESFET	NA	2.5	20	NA	NA	Derivative superposition	Fully integrated NA	Single ended
[25]	InGaP GaAs HBT	7.5	11	7	22	NA	Optimum bias condition	Fully integrated NA	Single ended
[28]	CMOS 0.18 μm	7	7.5	5	19	158	Multi-Tanh	Fully integrated 2.4 mm^2	Fully differential
[97]	CMOS 0.18 μm	8.4	3.7–8.8	11	19	154	MGTR	Fully integrated 2.5 mm^2	Single ended
This work [30]	CMOS 0.13 μm	5	9.5	20	18.5	128	Cross-coupled compensator	Fully integrated 1.5 mm^2	Fully differential bidirectional

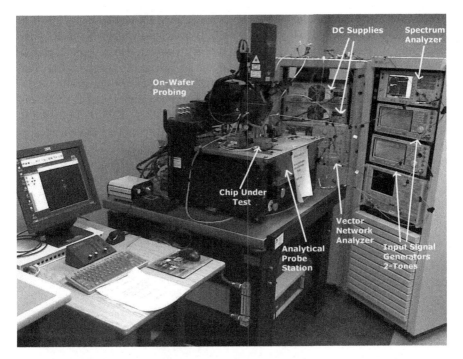

Fig. 7.18 Measurement setup of the 0.13 μm RF CMOS three-stage fully-differential linearized CMOS distributed bidirectional amplifier on analytical probe station

in the Fig. 7.18. The testing instruments also includes an Agilent RF spectrum analyzer for characterizing frequency content, a vector network analyzer, modulation analyzers and 8-pin high-frequency wafer probes.

7.5 Summary

In this chapter, the experimental measurement test-setups and measurement results for the fully-integrated fully-differential linearized CMOS distributed bidirectional amplifier are presented. The linearized distributed bidirectional amplifier achieves large 20 dB IMD3 distortion reduction over ultra-wideband frequency range for both RF paths with least power consumption and minimum silicon chip area solution compared to other published linearized DAs. The proposed fully-integrated DA linearization technique greatly suppresses the third-order intermodulation (IM3) distortion with drain and gate transmission-lines staggered to filter out the IM3 distortion. The proposed fully-differential linearized DA employs a CMOS cross-coupled compensator to enhance the linearity of the DA gain cell with a nonlinear drain capacitance compensator for wider linearization bandwidth. The proposed linearized CMOS bidirectional DA achieves a measured IM3 reduction of 20 dB in both RF directions with a two-way gain of 5 dB over ultra-wideband 0.1–9.5 GHz frequency of operation eliminating the need of RF switches which degrade performance and increase insertion loss. The proposed linearized DA is fully-differential suppressing substrate noise thus providing better dynamic range compared to single-ended linearized DA designs and with the least power consumption of 128 mW and chip silicon area compared to previous published work. An IIP3 of 18.5 dBm is achieved for both RF paths with a 10 dB IIP3 improvement. It is implemented in 0.13 μm RF CMOS technology and with a silicon chip area of 1.5 mm^2 for use in highly-linear low cost ultra-wideband communications.

Chapter 8
Conclusion

8.1 Summary

The emphasis on higher data-rates has driven the industry towards linear modulation techniques such as QPSK, QAM and multi-carrier configurations. Spectral efficiency has become a significant factor in the use of such linear modulation techniques. The result is a signal with a fluctuating envelope which generates intermodulation distortion from the power amplifiers. Linear modulation techniques are more spectral efficient however requires a linear power amplifier. Broadband power amplifiers are important RF components in a wireless communications system. All power amplifiers exhibit inherent nonlinearity which causes spectral regrowth in systems using non-constant envelope digital modulation schemes. The main source of spectral regrowth is the intermodulation distortion of the modulated carrier by nonlinearities in the transmitter power amplifier. Requirements for suppression of spectral regrowth have become more stringent and the improvement in spectral regrowth suppression is the primary reason for using power amplifier linearization techniques.

Active MOSFET devices can be modeled by nonlinear current sources that depend on the device voltages [12–14]. These nonlinear sources will give rise to distortion when driven with a modulated signal. The real MOSFET device output impedance is nonlinear and the mobility μ is not a constant but a function of the vertical and horizontal electric field. We may bias the active MOSFET device where the device behavior is nonlinear.

In this book, we demonstrated a fully-integrated fully-differential linearized CMOS distributed bidirectional amplifier that achieves large 20 dB IMD3 distortion reduction over ultra-wideband frequency range for both RF paths with least power consumption and minimum silicon chip area solution compared to other published linearized DAs. The proposed fully-integrated DA linearization technique greatly suppresses the third-order intermodulation (IM3) distortion with drain and gate transmission-lines stagger-compensated. Reducing and filtering out the DA IM3 distortion by mismatching the gate and drain LC delay-line ladder's time-delay.

The proposed fully-differential linearized DA employs a CMOS cross-coupled compensator to enhance the linearity of the DA gain cell with a nonlinear drain capacitance compensator for wider linearization bandwidth. The proposed linearized CMOS bidirectional DA achieves a measured IM3 reduction of 20 dB in both RF directions with a two-way gain of 5 dB over ultra-wideband 0.1–9.5 GHz frequency of operation eliminating the need of RF switches which degrade performance and increase insertion loss. The proposed linearized DA is fully-differential suppressing substrate noise thus providing better dynamic range compared to single-ended linearized DA designs and with least power consumption of 128 mW and chip silicon area compared to previous published work. An IIP3 of 18.5 dBm is achieved for both RF paths with a 10 dB IIP3 improvement. It is implemented in 0.13 μm RF CMOS technology with a silicon chip area of 1.5 mm^2 for use in highly-linear low cost bidirectional ultra-wideband communications. Comparing the proposed CMOS linearized DA in this work [30] to other published ones, the proposed linearized CMOS DA offers a 20 dB IM3 distortion reduction with 9.5 GHz operational bandwidth and with the least power consumption [30].

The book objectives were introduced in Chap. 1. In Chap. 2, modulation schemes effect on RF power amplifier nonlinearity and RFPA linearization techniques were presented. In Chap. 3, distributed amplification principles and transconductor nonlinearity compensation were presented. Various applications of linearized distributed circuit functions were presented in Chap. 4. Chapter 5 described in details the proposed fully-integrated linearized CMOS bidirectional distributed amplifier and the proposed highly-linear CMOS cross-coupled compensator transconductor with enhanced tunability. Chapter 6 presented the proposed linearized CMOS bidirectional distributed amplifier layout techniques and considerations. Chapter 7 presented the proposed linearized CMOS bidirectional distributed amplifier experimental test setups and measured results. Chapter 8 drew conclusions of the research work and listed research contributions.

8.2 Book Research List of Contributions

In summary, the main objective of this book research was aimed at the broadband amplifier linearization for providing broadband linear amplification for use in future Radio-over-Fiber communication and wireless transceiver applications. The following is a summary of the research contributions:

1. The development of linearized CMOS stagger-compensated bidirectional distributed amplifier with 20 dB IM3 distortion reduction.
2. The development of a highly-linear CMOS cross-coupled compensator transconductor with enhanced tunability and its theoretical analysis.
3. The application of linearization techniques to distributed circuit designs such as power splitters, matrix amplifier and paraphase amplifier.

8.3 Research Future Work

There are a few interesting new ideas that await exploration in future research of fully-integrated adaptive RF power amplifier linearizer modules for multi-carrier modulation schemes such as OFDM. Nonlinear signal distortions are generated since most transmitters operate their power amplifiers near saturation to achieve maximum power efficiency. This nonlinear distortion generates spurious spectral sidelobes and spreads the transmitted signal into the adjacent channel causing interference. To resolve this issue, advanced power amplifier linearization modules employing filters will be implemented in the power amplifier optimized transconductor linearization to suppress the sideband lobes.

Future research work will focus on the development of new generation of fully-integrated CMOS linearized broadband power amplifier modules. These advanced linearized power amplifier modules will include build-in optimized transconductors with digitally enhanced distortion compensation system-on-chip. Signal conditioning adaptive linearization circuits will be investigated that allow broadband power amplifiers to run closer to compression (saturation) with improved efficiency and reduced spectral regrowth for linear modulation techniques.

The work presented in this book presents various integrated circuit techniques applied to power amplifiers for spectral regrowth suppression and distortion cancellation. There are a number of improvements that can be made to the proposed linearized bidirectional distributed amplifier such as adaptive digital predistortion circuitry. The added circuitry will assist in maintaining a dynamically updated model of the broadband power amplifier optimizing distortion cancellation. Perform improvements to increase the circuit bandwidth by developing optimized transconductors with reduced gate and drain parasitic capacitance.

Appendix A
Quadrature Signal Processing

A.1 Analog Modulation Transmission

Signal modulation involves changes made to signal waves such as sine waves in order to encode information. There are two types of modulation transmission processes, analog and digital modulation.

In analog modulation, the modulation can be achieved by various methods such as amplitude modulation (analog AM), frequency modulation (analog FM) and phase modulation (analog PM) [31, 33]. Amplitude modulation is the process of changing the amplitude of a high frequency carrier signal with an analog signal modulating signal (information).

A.2 Why Digital Modulation Transmission?

Modern communications systems demand more information capacity and higher signal quality. AM and FM modulation methods do not meet the needs for higher data rate traffic. In digital modulation, an analog carrier signal is modulated by a digital bit stream (0 and 1). Once the baseband signal has been digitized, it is then transferred over an analog passband channel. Digital modulation can be achieved by various methods such as phase-shift keying (digital PSK), frequency-shift keying (digital FSK) and amplitude-shift keying (digital ASK) [31–33].

In advanced phase shift keying modulation signals are represented in phase domain. Phase domain uses a vector (or a point) in the plane to represent the signal. The length of the vector is the amplitude and the angle represents the phase. Higher order phase modulation schemes such as QPSK (Quadrature Phase Shift Keying) are often used when improved spectral efficiency is required. We can improve on information density by shifting between four states with each symbol having two bits of data. QPSK is capable of processing two bits for each symbol. QPSK has a higher tolerance level for channel link degradation [31, 32]. This is due to the

Z. El-Khatib et al., *Distributed CMOS Bidirectional Amplifiers: Broadbanding and Linearization Techniques*, Analog Circuits and Signal Processing, DOI 10.1007/978-1-4614-0272-5, © Springer Science+Business Media New York 2012

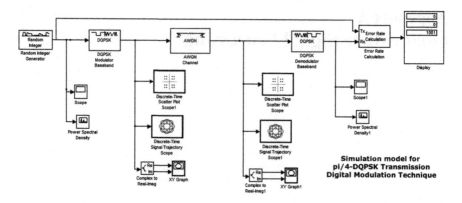

Fig. A.1 Simulink model test bench schematic for quadrature $\Pi/4$-DQPSK modulator and demodulator transmission

fact that QPSK has four possible states. QPSK is implemented in transmitters and receivers used by Wireless LAN communication systems. QPSK enables the system to efficiently utilize its bandwidth resources.

QPSK uses four points on the constellation diagram equispaced. In the case when the phase shift is 180°, the modulated carrier is turned off briefly as the I and Q values go through the origin of the I/Q plane. The transmitter output power amplifier must be linear over a wide dynamic range to prevent spectral splatter and therefore interference with adjacent frequency channels [32, 33].

To avoid the problem of splatter, cellular telephone systems such as NADC-TDMA uses $\Pi/4$-DQPSK [31–33]. A second set of four possible phase states is utilized with one set is on the I/Q axes and the second is offset by 45°. Differential means that the information is carried by the transition between states. In $\Pi/4$-DQPSK the set of constellation points are toggled each symbol, so transitions through the origin does not occur therefore less spectral splatter compared to QPSK. This scheme produces the lowest envelope variations. An extremely attractive feature of $\Pi/4$-DQPSK is that it can be non-coherently detected, which greatly simplifies receiver design. A noisy-channel can distort the constellation of a transmitted communication signal as shown in Fig. A.1. Simulink model test bench schematic for quadrature 16-QAM modulator and demodulator transmission is shown in Fig. A.1. Figure A.2 shows the simulink model test bench constellation output of a $\Pi/4$-DQPSK signal and its distorted version. Transmitter nonlinearity produces harmonics and intermodulation distortion (IMD) products. Some of these products fall within the transmission band and can degrade the system performance. Nonlinear distortions generate in-band interferences which results in amplitude and phase deviation of the modulated vector signal as shown in Fig. A.2.

Higher data rate communication requires large bandwidth and higher order constellation. QAM (Quadrature Amplitude Modulation) is another advanced modulation technique used in modern communications systems due to its bandwidth efficiency. Many fixed wireless radio systems use QAM for broadband applications. In order to increase the data rate in communication systems, the modulation scheme

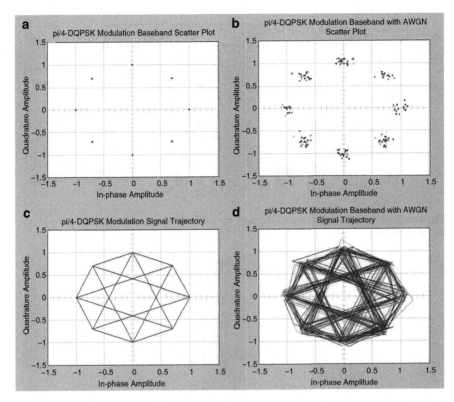

Fig. A.2 Simulink model test bench constellation output of a $\Pi/4$-DQPSK signal and its distorted version

can incorporate higher order constellations, however at the cost of tighter transmitter linearity requirement. It is therefore desirable to develop linearization techniques for RF power amplifiers transmitters.

Like all modulation schemes, QAM conveys data by combining amplitude and phase modulation. QAM takes advantage of the fact that greater number of symbols achieves greater system efficiency. Occupied bandwidth is determined by the symbol rate so the more bits (fundamental information units) per symbol thus achieving higher efficiency. It is possible to code n bits using one symbol. Discrete levels n equal to 2 is identical to QPSK. A 16-QAM corresponds to 4 bits equals one symbol. To transmit a symbol a modulator adjusts the vector to the point in signal space that corresponds to the desired symbol. A noisy-channel can distort the constellation of a transmitted communication signal as shown in Fig. A.3. Simulink model test bench schematic for quadrature 16-QAM modulator and demodulator transmission is shown in Fig. A.3. Figure A.4 shows the simulink model test bench constellation output of a 16-QAM signal and its distorted version. Transmitter nonlinearity produces harmonics and intermodulation distortion (IMD) products. Some of these products fall within the transmission band and can degrade the system

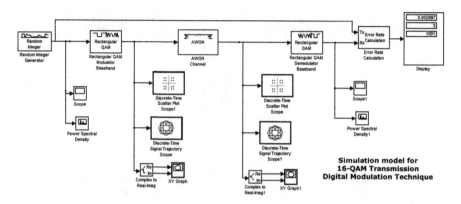

Fig. A.3 Simulink model test bench schematic for quadrature 16-QAM modulator and demodulator transmission

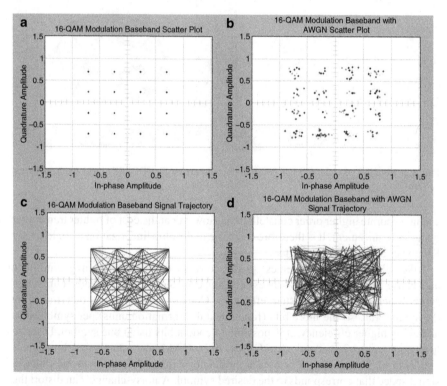

Fig. A.4 Simulink model test bench constellation output of a 16-QAM signal and its distorted version

performance. Nonlinear distortions generate in-band interferences which results in amplitude and phase deviation of the modulated vector signal as shown in Fig. A.4.

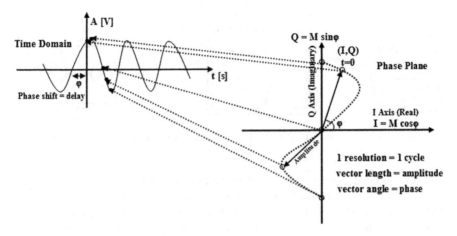

Fig. A.5   Time domain transmitted signal mapping into phase plane

A.2.1   Time Domain Transmitted Signal Mapping into Phase Plane

The transmitted signal can be viewed in different domains. Signal performance is usually analyzed in time or frequency domain. However, for devices operating on a signal's phase, the phase domain becomes useful. A single sine wave cycle can be mapped from the time domain into phase plane as shown in Fig. A.5 [31–33]. Changes in phase in the time domain show up as changes in location in the phase plane. Where the signal angle corresponds to the location in the cycle and the length corresponds to the signal strength [31, 33].

A.2.2   IQ Data Modulation Technique in Quadrature Processing Systems

The function of a modulator is to add data to a carrier by altering the carrier's amplitude, frequency and phase in a controlled manner [33, 34]. The simplest modulators are VCOs (FM) and variable amplifiers (AM) as shown in Fig. A.6 [34]. By altering the output power in a controlled manner, an amplifier becomes an amplitude modulator. Phase modulation is achieved using a mixer component. The mixer is used for a bi-phase (two-state) modulator [34]. By summing data and carrier, that is by multiplying the carrier by 1 and −1, a mixer becomes a bi-phase modulator.

For QPSK modulation we need two mixers and a summer [33]. This allows the creation of any point in the phase plane with one mixer for the I axis and one for Q. The data is split into two streams called I (in phase) and Q (quadrature). These

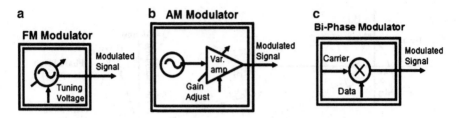

Fig. A.6 Basic modulators: (a) FM modulator (b) AM modulator (c) Bi-phase modulator

data streams need to be offset by 90°. By summing two orthogonal data streams and a carrier, we get a phase modulator producing any angle in the phase plane. If we add amplitude control we get a vector modulator producing any point in the phase plane. IQ modulation technique is an efficient way to transfer information and it also works well with digital formats in RF communications systems [34]. When we modulate a carrier with a waveform that changes the carrier's frequency slightly, we can treat the modulating signal as a phasor. It has both a real and an imaginary part. If we build a receiver that locks to the carrier then we can retrieve information by reading the I and Q parts of the modulating signal. The information appears on a polar plot. The I/Q plane shows us what the modulated carrier is doing relative to the unmodulated carrier and what baseband I and Q inputs are required to produce the modulated carrier [124].

For a circuit that uses I and Q waveforms it is difficult to precisely vary the phase of a high frequency carrier sine wave in a hardware circuit according to an input message signal [33]. Both the flexibility and simplicity of the design of an IQ modulator are the main reasons for its widespread use and popularity. A hardware signal modulator that manipulated the magnitude and phase of a carrier sine wave would be expensive and difficult to design. To understand how we can avoid having to manipulate the phase of an RF carrier directly we first return to trigonometry.

The three fundamental parts of a RF signal are amplitude, frequency and phase [41]

$$A \cos (\omega_c t + \phi) = A \cos (\omega_c t) \cos (\phi) - A \sin (\omega_c t) \sin (\phi) \qquad (A.1)$$

$$I = A \cos (\phi) \qquad (A.2)$$

$$Q = A \sin (\phi) \qquad (A.3)$$

$$A \cos (\omega_c t + \phi) = I \cos (\omega_c t) - Q \sin (\omega_c t) \qquad (A.4)$$

where I is the amplitude of the in-phase carrier and Q is he amplitude of the quadrature phase carrier.

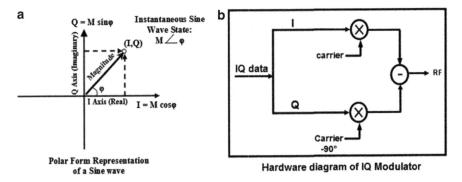

Fig. A.7   (**a**) Polar form representation of a sine wave. (**b**) Hardware diagram of IQ data modulator

Figure A.7b shows a block diagram of an IQ data modulator [32,33]. I and Q data are used to represent any changes in magnitude and phase of a message signal. Since the IQ modulator is basically reacting to changes in I and Q waveform amplitudes it can be used for any modulation scheme. This technique is known as quadrature up-conversion [33]. The mixer device performs frequency multiplication and either up-convert or down-convert signals. The IQ modulator mixes the I waveform with the RF carrier sine wave and mixes the Q signal with the same RF carrier sine wave offset by 90° in phase [32,33].

The Q signal is subtracted from the I signal as in (A.1) producing the final RF modulated waveform. The shifting of the carrier by 90° is the source of the names for the I and Q data. I refers to in-phase data (since the carrier is in phase) and Q refers to quadrature data (since the carrier is offset by 90°). I/Q modulator output is plotted on the I/Q plane as vector phasors and symbols with Sine on Q axis and Cosine on I axis as shown in Fig. A.7 [32, 33]. A cosine wave and a sine wave with the same frequency are similar except for a 90° phase offset between them. The is very important since it means that we can control the amplitude, frequency and phase of a modulating RF carrier sine wave by manipulating the amplitudes of separate I and Q input signals. This way we don't have to directly vary the phase of an RF carrier sine wave instead we can achieve the same effect by manipulating the amplitudes of input I and Q signals. A 90° phase shifter hardware circuit is included between the carrier signals used for the I and Q mixers which is a simpler simpler design issue than the direct phase manipulation [32,33].

References

1. Hittite Microwave (2010) High IP3 mixers for cellular applications. Hittite product application notes. Hittite, Massachusetts, USA
2. Larson L, Asbeck P (2008) Advanced digital linearization approaches for wireless RF power amplifiers. In: IEEE circuits and systems workshop: system-on-chip – design, applications, integration and software, San Diego, CA, USA, pp 1–7
3. Smith D et al (1997) 480 Mbps ultra-wideband radio-over-fibre transmission using a 1310/1550 nm reflective electro-absorption transducer and off-the-shelf components. IEEE J. Solid-State Circuit, pp 1–17
4. Thakur MP, Ben-Ezra Y (2009) 480 Mbps, Bi-directional, ultra-wideband radio-over-fiber transmission using a 1308/1564 nm reflective electro-absorption transducer and commercially available VCSELs. J Lightwave Technol 27(3):266–272
5. Fernando XN, Sesay AB (2000) Higher order adaptive filter based predistortion for nonlinear distortion compensation of radio over fiber links. IEEE Int Conf Commun 1:367–371
6. Sadhwani R et al (2003) Adaptive CMOS predistortion linearizer for fiber-optic links. J Lightwave Technol 21(12):3180–3193
7. Black HS (1984) Stabilized feed-back amplifiers. Proc IEEE 72(6):716–722
8. Ghoniemy S, MacEachern L, Mahmoud SA (2003) Performance analysis and enhancement of RF/Fiber optical interface for microcellular wireless transceivers. In: Proceedings of wireless and optical communications, Banff
9. Kim TW et al (2004) Highly linear receiver front-end adopting MOSFET transconductance linearization by multiple gated transistors. IEEE J Solid-State Circuit 39(1):223–229
10. Tsun Mok K et al (2004) Linearised distributed amplifier with low linearisation loss. Electron Lett 40(1):2011–2014
11. Raab F, Sokal N (2002) Power amplifiers and transmitters for RF and microwave. IEEE Trans Microw Theory Tech 50:814–826
12. de Vreede L, van der Heijden M (2006) Linearization Techniques at the Device and Circuit Level. IEEE BCTM, pp 1–8
13. Wambacq P, Sansen W (1998) Distortion analysis of analog integrated circuits. Kluwer, Boston
14. Maas S (1988) Nonlinear microwave circuits. Artech House, Norwood
15. Mbabele M, Aitchison CS (2001) Third order intermodulation improvement in distributed amplifiers. In: IEEE European microwave conference, Microwave Engineering Europe, London, pp 1–4
16. Minasian RA (1980) Intermodulation distortion analysis of MESFET amplifiers using the volterra series representation. IEEE Trans Microw Theory Tech 28:1–8

Z. El-Khatib et al., *Distributed CMOS Bidirectional Amplifiers: Broadbanding and Linearization Techniques*, Analog Circuits and Signal Processing, DOI 10.1007/978-1-4614-0272-5, © Springer Science+Business Media New York 2012

17. Paul DK, Parkinson G (2005) A New Approach for the Linearization of a Distributed Amplifier. Microwave and Optical Technology Letter 46(1):15–17
18. Lee CH et al (2004) Enhanced performance of ROF link for cellular mobile systems using postdistortion compensation. In: Proceedings of the IEEE international symposium on personal, indoor and mobile radio communications, IEEE, Piscataway
19. Leisten OP et al (1988) Distributed amplifiers as duplexer/low crosstalk bidirectional elements in S-Band. Electron Lett 24:188–189
20. Byrne JW, Beyer JB (1989) A highly directive, broadband, bidirectional distributed amplifier. IEEE Int Microw Symp Dig 1:188–189
21. Sundaram B, Prasad PN (2006) A novel electronically tunable active duplexer for wireless transceiver applications. IEEE Trans Microw Theory Tech 54:2584–2592
22. Cryan MJ et al (2000) Analysis and design of integrated active circulator antennas. IEEE Trans Microw Theory Tech 48(6):1017–1023
23. Lau K, Chan C (2008) Self-biased distributed amplifier: linearity improvement and efficiency enhancement. Microw Opt Technol Lett 50(10):2493–2497
24. Tsun Mok K et al (1996) Control of circuit distortion by the derivative superposition method. In: IEE colloquium wideband circuits, modelling and technique, Savoy Place, pp 2011–2014
25. Iwamoto M, Hutchinson CP (2002) Optimum bias conditions for linear broadband In-GaP/GaAs HBT power amplifiers. IEEE Int Microw Symp Dig 2:901–904
26. Gilbert B (1998) The multi-tanh principle: a tutorial overview. IEEE J Solid-State Circuit 33:188–189
27. Lu C, Pham AH, Livezey D (2006) On the feasibility of CMOS multiband phase shifters for multiple-antenna transmitters. IEEE Microw Wirel Compon Lett 3:2100
28. El-Khatib Z, MacEachern L, Mahmoud SA (2009) A fully-integrated linearized CMOS distributed amplifier based on multi-tanh principle for radio over fiber and ultra-wideband applications. IEEE Radio Wirel Symp 33:188–189
29. Shiroma G, Miyamoto RY, Shiroma WA (2005) A combined distributed amplifier/true-time-delay phase shifter for broadband self-steering arrays. IEEE MTT-S Int Microw Symp Dig 16:4
30. El-Khatib Z, MacEachern L, Mahmoud SA (2010) Linearised bidirectional distributed amplifier with 20 dB IM3 distortion reduction. Electron Lett 46(15):1089
31. Stern HPE, Mahmoud SA, Stern LE (2004) Communication systems: analysis and design. Pearson Prentice Hall, Upper Saddle River
32. Rogers J, Plett C (2010) Radio frequency integrated circuit design. Artech House, Boston
33. Razavi B (1998) RF microelectronics. Prentice Hall, Englewood Cliffs
34. Lee TH (2004) The design of CMOS radio frequency integrated circuits. Cambridge University Press, Cambridge/New York, pp 423–425
35. Webster DR, Parker A (1996) Derivative superposition – a linearization technique for ultra broadband systems. In: IEEE colloquium wideband circuits, modelling and techniques. IEEE, London
36. Vuolevi J, Rahkonen T (2003) Distortion in RF power amplifiers. Artech House, Boston
37. Gonzalez G (1997) Microwave transistor amplifiers: analysis and design. Prentice Hall, Upper Saddle River
38. Steer M, Larson LE (2005) The impact of RF front-end characteristics on the spectral regrowth of communications signals. IEEE Trans Microw Theory Tech 53(6):2179
39. Zhou G (2000) Analysis of spectral regrowth of weakly nonlinear power amplifiers. IEEE Int Conf Acoust Speech Signal Process 5:2737–2740
40. Bahl IJ (2009) Fundamentals of RF and microwave transistor amplifiers. Wiley, Hoboken, pp 40–43
41. Wu Q, Larkin R (1996) Linear RF power amplifier design for CDMA signals. IEEE MTT-S Int Microw Symp Dig 2:851
42. Sevic J, Staudinger J (1996) Simulation of adjacent-channel power for digital wireless communication systems. IEEE MTT-S Int Microw Symp Dig 2:851–854

43. Carvalho NB, Pedro JC (1999) Compact formulas to relate ACPR and NPR to two-tone IMR and IP3. Microw J 42(12):70–85
44. Struble W et al (1997) Understanding linearity in wireless communication amplifiers. IEEE J Solid-State Circuit 32(9):1310
45. Carvalho NB, Pedro JC (2003) Intermodulation distortion in microwave and wireless circuits. Artect House, Boston
46. Heutmaker MS (2003) Error vector and power amplifier distortion. In: Proceedings of the 1997 annual wireless communications, IEEE, Piscataway
47. Petrovic V, Gosling W (1979) Polar-loop transmitter. Electron Lett 15(10):286–287
48. Briffa MA, Faulkner M (1996) Stability analysis of Cartesian feedback linearisation for amplifiers with weak nonlinearities. IEE Proc Commun 143(4):212–218
49. Kenington P (2000) High-linearity RF amplifier design. Artech House, Boston
50. Hittite Microwave (1994) Linearisation of RF multicarrier amplifiers using Cartesian feedback. Electron Lett 45:1110–1111
51. Petrovic V, Gosling W (1979) Polar-loop transmitter. Electron Lett 15(10):286–288
52. Cripps S (2002) Advanced techniques in RF power amplifier design. Artech House, Boston
53. Cavers JK (1990) Amplifier linearization using a digital predistorter with fast adaptation and low memory requirements. IEEE Trans Veh Technol 39(4):374–382
54. Nezami MK (2004) Fundamentals of power amplifier linearization using digital predistortion. High Freq Des Electron 3(8):54–59
55. Chen Y, Beyer JB, Sokolov V, Culp J (1986) A 11 GHz hybrid Paraphase amplifier. IEEE Int Solid-State Circuit Conf XXIX:236–237
56. Ginzton EL, Hewlett WR, Jasberg JH, Noe JD (1948) Distributed amplification. Proc IRE 36:956–969
57. Wong TTY (1993) Fundamentals of distributed amplification. Artech House, Boston
58. Razavi B (2001) Design of analog CMOS integrated circuits. McGraw-Hill, New York
59. Razavi B (2003) Design of integrated circuits for optical communications. McGraw-Hill, New York
60. Beyer JB et al (1984) MESFET distributed amplifier design guidelines. IEEE Trans Microw Theory Tech 32(3):268–275
61. Prasad JB et al (1988) Power-bandwidth considerations in the design of MESFET distributed amplifiers. IEEE Trans Microw Theory Tech 36(7):1117–1123
62. Ayasli Y et al (1982) A monolithic GaAs 1-13-GHz traveling-wave amplifier. IEEE Trans Microw Theory Tech 30(7):976–981
63. Ahn H, Allstot D (2002) A 0.5–8.5 GHz fully differential CMOS distributed amplifier. IEEE Microw Theory Tech 43(6):985–993
64. Lu C, Pham AH, Livezey D (2006) On the linearity of CMOS multi-band phase shifters. In: IEEE silicon monolithic integrated circuits in RF systems digest, vol 1. IEEE, Piscataway, p 4
65. Salama CAT (2003) A novel C-band CMOS phase shifter for communication systems. IEEE Int Symp Circuit Syst 2:316
66. Lu C, Pham AH, Livezey D (2005) A novel multiband phase shifter with loss compensation in 180 nm RF CMOS technology. In: IEEE midwest symposium on circuits and systems, vol 1. Lansing, p 806
67. Pozar DM (1990) Microwave engineering. Addison Wesley Publishing Company, Reading
68. Sarma DG (1954) On distributed amplification. Proc Inst Elect Eng 102B:689-697
69. Deibele S et al (1989) Attenuation compensation in distributed amplifier design. IEEE Trans Microw Theory Tech 37(9):1425-1433
70. Popplewell P, Karam V, Shamim A et al (2008) A 5.2-GHz BFSK transceiver using injection-locking and an on-chip antenna. IEEE J Solid-State Circuit 43:6
71. Kenington PB, Wilkinson RJ, Marvill JD (1991) Broadband linear amplifier design for a PCN base-station. In: IEEE vehicular technology conference, St. Louis, pp 155–160
72. Kimura K (1997) The ultra-multi-tanh technique for bipolar linear transconductance amplifiers. IEEE Trans Circuit Syst 44(4):288

73. Li ZM, Prasad PN (1996) Optimal design of low crosstalk, wideband, bidirectional distributed amplifiers. IEEE Int Microw Symp Dig 2:847
74. Percival WS (1937) Thermionic valve circuits. British Patent Specification no. 460562
75. Lu C, Pham AH, Livezey D (2006) Development of multiband phase shifters in 180-nm RF CMOS technology with active loss compensation. IEEE Trans Microw Theory Tech 54:40–45
76. Safarian A, Heydari P (2007) CMOS distributed active power combiners and splitters for multi-antenna UWB beamforming transceivers. IEEE J Solid-State Circuit 42(7):1481–1491
77. Hilder GM (1991) A hybrid 2–40 GHz high power distributed amplifier using an active splitter and combined drain line configuration. In: IEE colloquium millimetre wave transistors and circuits, Institution of Electrical Engineers, London
78. El-Khatib Z, MacEachern L, Mahmoud SA (2009) CMOS distributed paraphase amplifier employing derivative superposition linearization for wireless communications. In: IEEE midwest symposium on circuits and systems, Cancun
79. El-Khatib Z, MacEachern L, Mahmoud SA (2009) CMOS distributed active power splitter with multiple-gated transistor linearization for ultra-wideband applications. In: IEEE microsystems and nanoelectronics research conference, Ottawa
80. Pan HM, Larson LE (2007) Highly linear bipolar transconductor for broadband high-frequency applications with improved input voltage swing. IEEE Trans Microw Theory Tech 49 pp 713–716
81. Kim B, Ko J, Lee K (2000) A new linearization technique for MOSFET RF amplifier using multiple-gated transistor. IEE Microw Guid Wave Lett 10(9):371–373
82. Tanaka S, Abidi AA (1997) A linearization technique for CMOS RF power amplifiers. In: IEEE symposium VLSI circuits digest technical papers, Kyoto
83. Kang S, Kim B (2003) Linearity analysis of CMOS for RF applications. IEEE Trans Microw Theory Tech 51(3):972–977
84. El-Khatib Z, MacEachern L, Mahmoud SA (2004) A 0.1–12 GHz fully differential CMOS distributed amplifier employing a feedforward distortion cancellation technique. IEEE Int Symp Circuit Syst 1:I-617–I-620
85. Popplewell P, Karam V, Shamim A et al (2008) A 5.2-GHz BFSK transceiver using injection-locking and an on-chip antenna. IEEE J Solid-State Circuit 43:981–990
86. Shamim A (2006) Silicon differential antenna/inductor for short range wireless communication applications. In: Canadian conference on electrical and computer engineering, Ottawa
87. Chen T, Chien J, Lu L (2005) A 45.6-GHz matrix distributed amplifier in 0.18 um CMOS. In: IEEE custom integrated circuits conference, San Jose
88. Chu S, Tajima Y (1989) A novel 4–18 GHz monolithic matrix distributed amplifier. In: IEEE MTT-S Digest, Long Beach
89. Niclas K (1991) Recent advances in the design of matrix amplifiers. Arch fur Elektrotech 74(3):225–229
90. Chen Y, Beyer JB, Sokolov V, Culp J (1991) Design and performance of 20 dB gain two-tier matrix distributed amplifer. Electron Lett 27:506–507
91. Kobayashi K, Esfandiari R, Streit D (1993) GaAs HBT wideband matrix distributed and darlington feedback amplifier to 24 GHz. IEEE Trans Microw Theory Tech 39:2001–2009
92. Park J, Allstot D (2004) A 12.5 GHz RF matrix amplifier in 180 nm SOI CMOS. IEEE Int Symp Circuit Syst 1:117–120
93. Gilbert B (1998) The multi-tanh principle: a tutorial overview. IEEE J Solid-State Circuit 33(1):2
94. Arapin V, Larson LE (2004) Linearization of CMOS LNAs via optimum gate biasing. IEEE Int Symp Circuit Syst 4:IV-748–IV-751
95. Chen YW, Beyer JB, Prasad SN (1985) MESFET wideband distributed paraphase amplifier. Int J Electron 58(4):553–569
96. Levent-Villegas M (1984) X-band paraphase amplifier. Electron Lett 20(11):451–453
97. Lu C, Pham AH (2007) Linearization of CMOS broadband power amplifiers through combined multigated transistors and capacitance compensation. IEEE Int Symp Circuit Syst 55 pp 2320–2328

98. El-Khatib Z, MacEachern L, Mahmoud SA (2009) Fully-integrated multi-band tunable linearized CMOS active analog phase shifter with active loss compensation for wireless home network multiple antenna transceiver applications. In: IEEE proceedings of the 2009 international symposium on circuits and systems, IEEE, Piscataway

99. El-Khatib Z, MacEachern L, Mahmoud SA (2010) Highly-linear CMOS cross-coupled compensator transconductor with enhanced tunability. Electron Lett 46(24):1597–1598

100. Webster D et al (1996) Control of circuit distortion by the derivative superposition method. IEEE Microw Guid Wave Lett 6(3):123–125

101. Toole B et al (2004) RF circuit implications of moderate inversion enhanced linear region in MOSFETs. IEEE Trans Circuit Syst 51(2):319–328

102. Quinn P (1981) A cascode amplifier nonlinearity correction technique. Int Solid-State Circuit Conf XXIV:188–189

103. Mbabele M, Aitchison CS (2001) Third order intermodulation improvement in distributed amplifiers. In: IEEE European microwave conference, Microwave Engineering Europe, London

104. Mbabele M, Aitchison CS (1980) Intermodulation distortion analysis of MESFET amplifiers using the volterra series representation. IEEE Trans Microw Theory Tech 28:1–8

105. Kenney J, Stevenson L (1995) Power amplifier spectral regrowth for digital cellular and PCS applications. Microw J 38:851–854

106. Quinn PA et al (1981) A cascode amplifier nonlinearity correction technique. IEEE Int Solid-State Circuit Conf XXIV:188–189

107. Allstot DJ (2005) A Full-range all-pass variable phase shifter for multiple antenna receivers. IEEE Trans Microw Theory Tech 16(3):2100–2103

108. Zolfaghari A, Chen A, Razavi B (2001) Stacked inductors and transformers in CMOS technology. IEEE J Solid-State Circuit 36(4):620–628

109. Dickson T, Voinigescu S et al (2005) 30–100-GHz inductors and transformers for millimeter-wave (Bi)CMOS integrated circuits. IEEE Trans Microw Theory Tech 53(1):123–133

110. Yue CP et al (1999) Design strategy of on-chip inductors for highly integrated RF systems. Des Autom Conf 982–987

111. Koutsoyannopoulos Y et al (2000) Performance limits of planar and multi-layer integrated inductors. In: IEEE international symposium on circuits and systems, Geneva

112. Burghartz JN et al (1998) Progress in RF inductors on silicon – understanding substrate losses. In: Proceedings of technical digest IEDM, San Francisco, pp 523-526

113. Gil J et al (2003) A simple wide-band on-chip inductor model for silicon-based RF ICs. IEEE Trans Microw Theory Tech 51(9):2023

114. Long J, Copland MA (1997) The modeling, characterization, and design of monolithic inductors for silicon RF IC's. IEEE J Solid-State Circuit 32:357–369

115. Yue CP, Wong SS (1998) On-chip spiral inductors with patterned ground shields for si-based RF ICs. IEEE J Solid-State Circuit 33:743–752

116. Razavi B (2002) Prospects of CMOS technology for high-speed optical communication circuits. IEEE J Solid-State Circuit 37(9):1135–1145

117. IBM Foundry and Manufacturing Services Education (2007) IBM CMOS8RF process. IBM CMOS8RF technology design kit manual. http://www.ibm.com/chips (Accessed on 27 Mar 2006)

118. Cho H et al (1991) A three-step method for the de-embedding of high frequency S-parameter measurements. IEEE Trans Electron Dev 38(6):1371-1375

119. Niknejad AM (2002) Modeling of passive elements with ASITIC. In: IEEE radio frequency integrated circuits symposium, IEEE, Piscataway

120. Kuhn WB et al (1998) Spiral inductor substrate loss modeling in silicon RF ICs. In: IEEE radio and wireless conference, IEEE, New York, pp 305–308

121. Koolen M (1992) Microelectronic engineering: on-wafer high-frequency device characterization. Elsevier Science Publisher, Eindhoven, The Netherlands

122. Agilent Technologies Agilent E4438C ESG vector signal generator data sheet. www.agilent.com/find/esg (Accessed on 14 Oct 2010)
123. Webster DR, Parker AE (1996) Derivative superposition – a linearisation technique for ultra broadband systems. In: IEE colloquium wideband circuits, modelling and techniques, Savoy Place, pp 901–904
124. Ellinger F (2008) Radio frequency integrated circuits and technologies. Springer, Berlin